DIVINE PROPORTION

Φ (Phi) In Art, Nature, and Science

Published by Sterling Publishing Co., Inc.
387 Park Avenue South, New York, NY 10016
by arrangement with The Book Laboratory ® Inc.

Distributed in Canada by Sterling Publishing
c/o Canadian Manda Group, 165 Dufferin Street
Toronto, Ontario, Canada M6K 3H6

For information about custom editions, special sales, premium and corporate purchases, please contact Sterling Special Sales Department at 800-805-5489 or specialsales@sterlingpub.com.

Editor: Rachel Federman

Design: Michael Zipkin

Art Research: Jeannine Jourdan

Figure Drawings: John Feld

ISBN 1-4027-3522-7

1 3 5 7 9 8 6 4 2

Manufactured in China

DIVINE PROPORTION

Φ (Phi) In Art, Nature, and Science

Priya Hemenway

STERLING PUBLISHING
NEW YORK

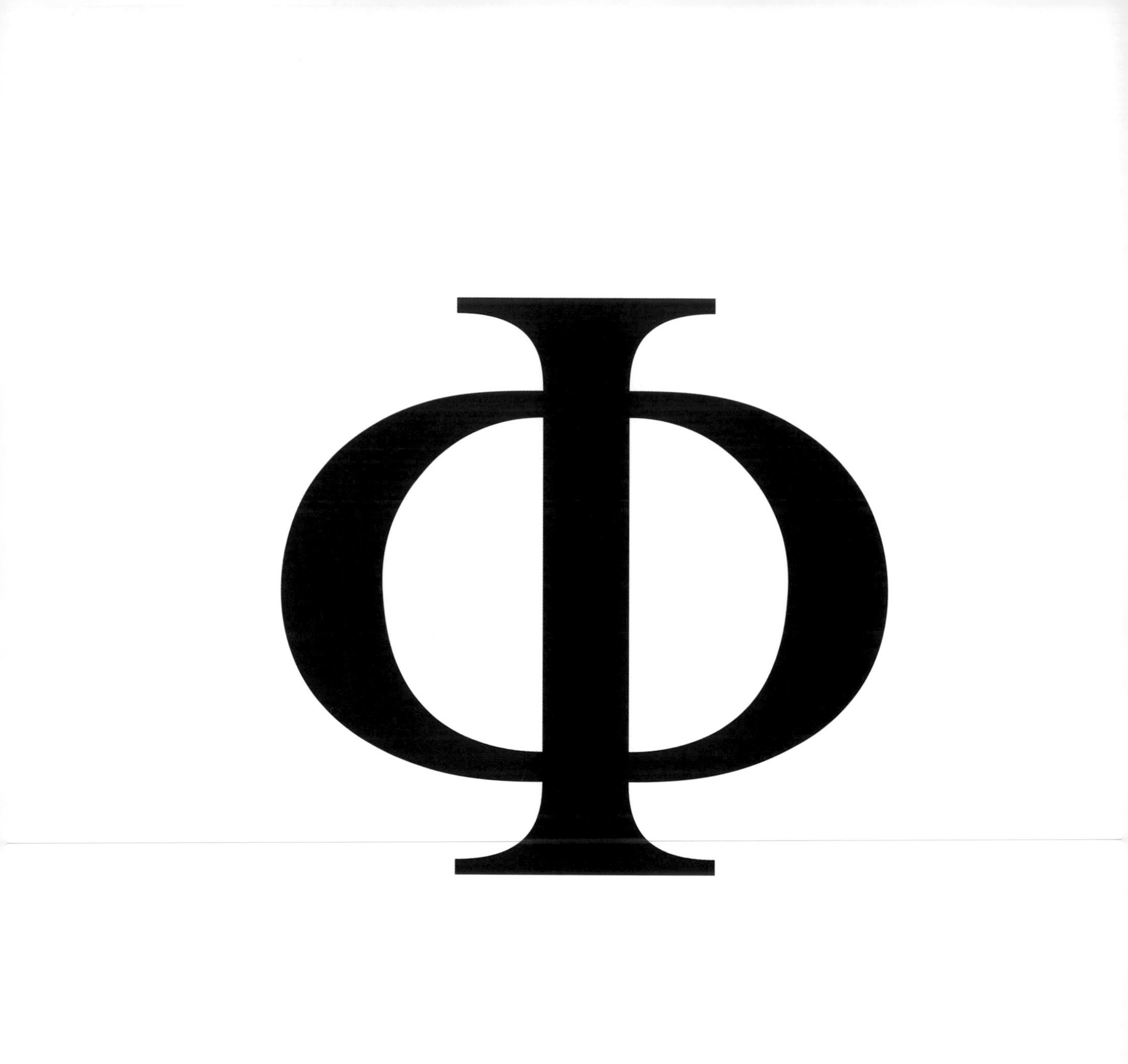

CONTENTS

INTRODUCTION

Man, know thyself in true proportion.
ORACLE OF DELPHI

SEEKERS OF TRUTH HAVE long been awed by the complexity of our universe. Physicists send probes into space, historians attempt to piece together fragments from our past, and botanists study nature's secrets. All agree that life, if nothing else, is endlessly mysterious. Gazing at a myriad of different forms and rhythms, they contemplate patterns, relationships, and telltale signs. How happy it would make us to come upon a clue or a formula that was a key to some unifying principle.

The spiral shape of the Chambered Nautilus (*Nautilus pompilius*) grows larger by a proportion of Φ—the Divine Proportion.

In proportion we find a strategy to do this; in Divine Proportion, a real possibility. Using the language of comparison and mathematical relationships we apply the Divine Proportion to life's mysteries by placing the larger next to the smaller and holding them both up to the whole. What we discover is a relationship of balance, harmony, and symmetry that is quite uncanny; and it is as mysterious in its functioning as the code we seek to break.

Not obvious, and not hidden, the Divine Proportion is easily expressed in words: The whole is to the larger in exactly the same proportion as the larger is to the smaller. It is as easily described as a pattern of numbers that increases by adding the

OPPOSITE:
Newton by William Blake (1757–1827)

To see a World in a Grain of Sand

And a Heaven in a Wild Flower,

Hold Infinity in the palm of your hand

And Eternity in an hour.

WILLIAM BLAKE

IN A VOYAGE that takes us from embryo to stars we will encounter the many faces of Φ. Sometimes the Divine Proportion can be mysteriously perplexing and enigmatic, leaving us wondering what it really is all about. Sometimes it is so obvious, we wonder that we had not noticed it before.

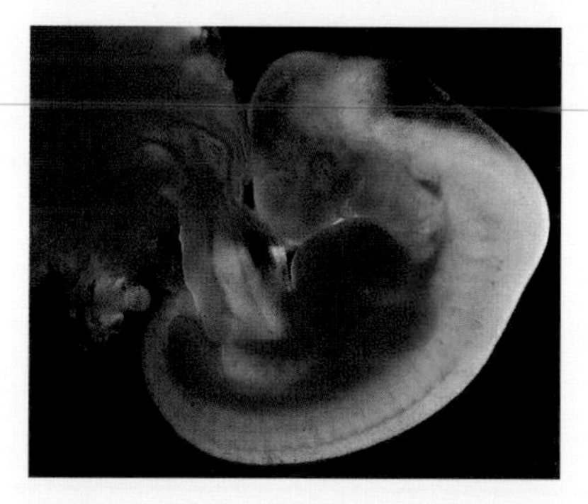

two previous numbers. But what does this mean? For those of us who find numbers and abstractions difficult to understand, it means there is a relationship that can be proven with numbers, that gives rise to a series of shapes and dynamics that appear throughout nature and which can be directly translated into rules of proportion to be used by artists. Furthermore, its principles of harmony have been acknowledged as fundamental truths in the world of spirit and its relationship to our daily lives is seen in the proportions of our very own bodies.

Human fascination with the Divine Proportion over many hundreds of years is to a great extent due to its many remarkable properties—harmony, regeneration, and balance are only a few. Its harmony is apparent in the principles of design that nature uses to give us patterns in plants, shells, the wind, and the stars. The regenerative principle shows up in shapes and solids that form the basis of everything from DNA to the contour of the universe. Balance is found in the spiral in our inner ear and is reflected in the unfurling shape of the human embryo that hurls us into existence.

In the everyday world of observing and measuring things, proportion is generally used to express the relation of parts to each other or to a whole and is built upon the similarity of two ratios. A ratio—simple in its structure—is a comparison. Terms like three-quarters or two hundred percent are ratios, and they are everywhere. We use them daily to compare all kinds of things.

To create a proportion we compare two ratios and introduce a term of measurement to establish how the comparison works.

When we want to pay closer attention to the exactness of the proportion (as mathematicians do) we use phrases to explain by how much one is proportional to the other.

> *Shall I compare thee to a summer's day?*
> *Thou art more lovely and more temperate...*
>
> SHAKESPEARE

> *Women have served all these centuries as looking-glasses possessing the magic and delicious power of reflecting the figure of a man at twice its natural size.*
>
> VIRGINIA WOOLF

For another example of the way that proportion works, we can look at what happens when we stand in front of a mirror. As we move closer we see less of ourselves, but the reflection is bigger; when we move away we see more, but the reflection is smaller. Thus we have two ratios—close is to big and far is to small—and a proportion—the size of the reflection is proportional to the distance from the mirror.

When we want to describe a ratio with numbers we use a line to divide its two parts. To describe a proportion we place an equal sign between two appropriate ratios. The mathematical description for the Divine Proportion is the Greek symbol Φ (Phi) and its algebraic formula says that $\Phi = \frac{(1+\sqrt{5})}{2}$.

Euclid, the famous Greek mathematician who first put the Divine Proportion into words, divided a line into two sections in such a way that the ratio of the whole line to the larger part is the same as the larger part is to the smaller.

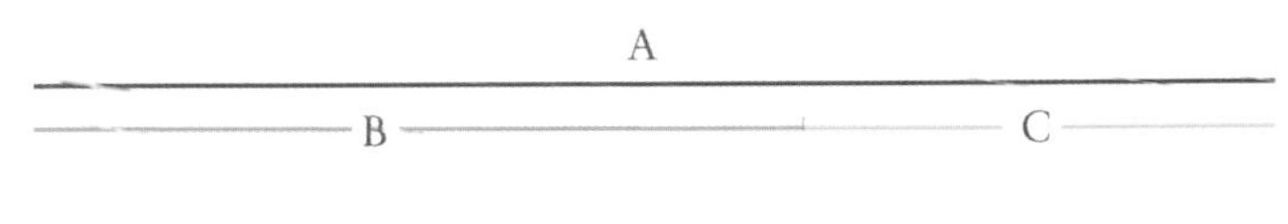

$$\frac{A}{B} = \frac{B}{C}$$

Translated into geometrical shapes this proportion, Φ, magically describes many of the patterns we see in nature. Used by architects it produces buildings of outstanding symmetry; extended into science it evokes abstract principles with application in other dimensions that are mind-bending. We see Φ expressed in the pyramids of Egypt, the Parthenon in Athens, and European Gothic cathedrals; we see Φ used by artists and artisans throughout the ages; and we see it as a perfect description for principles of growth and dynamism in nature.

History has provided us with a long list of names of people—mystics, philosophers, musicians, scientists, poets, artists, statesmen, mathematicians, lovers of wisdom, and ordinary people—who have experienced a collective fascination with the harmony of this wonderful proportion and who have probed its secrets to discover ever new and exciting properties. Century after century, from one discipline to another, virtues of the Divine Proportion have emerged and reemerged—appearing in ancient artifacts, in observations of nature, and in philosophy and mathematical wizardry. It has been wonderfully mirrored in art and architecture, echoed by music, and embraced by poets; all acknowledging its transcendental aspects. Why?

Looking out from inside the Parthenon in Athens we feel the strength and timeless beauty of its proportions.

The most essential and universally fascinating attraction of the Divine Proportion comes from the ratio's subliminal reference to ourselves, for its most profound aspect is explicitly evoked when we put the human being into the equation and perceive the proportion as a relationship that includes ourselves. Contemplating our place in the equation, we will variously find that we are the whole, we are the larger, and we are the smaller of the parts—always in some perfectly balanced ratio of whole to larger and larger to smaller.

As a relationship of macrocosm and microcosm, the Divine Proportion describes the larger and the smaller in their most intimate relationship: They are not separate; they are related. The proportion links them in such a way that there is a mirroring effect that makes it possible to see the large in the small and the small in the large.

Using language to help us look through the lenses of perception, we describe our human body as a universe, or see the "universe in a grain of sand." We can read a shape by its shadow, know a book by its cover, or understand a person's mood by the tone of their voice. We speak easily with similes, metaphors, and laws of simple proportion and find they make things clearer. The Divine Proportion provides us with a singularly unique lens with which to view our universe.

The remoter and more general aspects of the law are those which give it universal interest. It is through them that you not only become a great master in your calling, but connect your subject with the universe and catch an echo of the infinite, a glimpse of its unfathomable process, a hint of the universal law.

OLIVER WENDELL HOLMES, JR. (1841–1935), *THE PATH OF THE LAW*

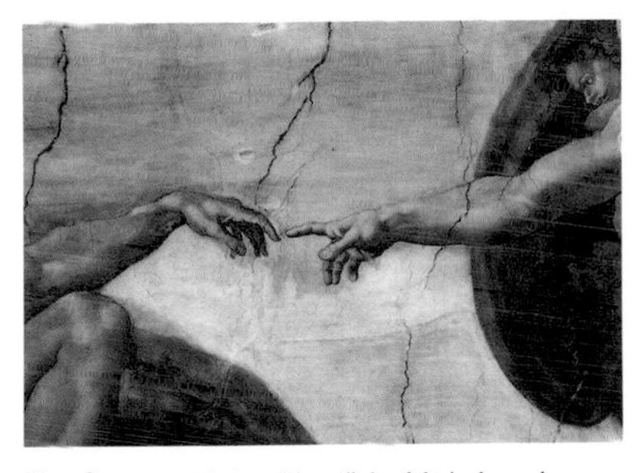

The Creation of Adam (detail) by Michelangelo (1475–1564) represents the divine moment when the fingers of God reach out and touch the fingers of Adam.

Universal laws are recognized in some deep part of ourselves as eternally true. To learn about the principles of Φ we will have to observe the many ways it is used. In a journey that will take us back in time to the building of the pyramids and forward into mystical expressions of the present, we will see the Divine Proportion unfold as a balanced and dynamic principle. We will learn different statements of its patterns—mathematical, creative, and spiritual—and we will experience its reflective nature. In all it is a fascinating study, for the many different manifestations of the Divine Proportion involve an understanding of times and places quite unlike our own. Like a musical sound with harmonious overtones, the Divine Proportion echoes an ancient wisdom, residing in a multitude of expressions; and, like all good things, this perfect proportion creates beauty.

There is nothing pleasurable except what is in harmony with the utmost depths of our divine nature.

HEINRICH SUSO (CA. 1295–1366), GERMAN MYSTIC

Chapter 1

Probing the Mysteries of Divine Proportion

Geometry has two great treasures: one is the theorem of Pythagoras; the other, the division of a line into extreme and mean ratio. The first we may compare to a measure of gold; the second we may name a precious jewel.

Johannes Kepler

Throughout the centuries many people have discovered, and sometimes rediscovered, the Divine Proportion. They were always impressed by its properties and have called it by a variety of different names. The Divine Proportion merits all the names it has been given: Divine Proportion, Golden Mean, Golden Proportion, Golden Section, Golden Ratio, Sacred Cut. All these descriptions refer to the proportion that is mathematically described as Φ (Phi). Simply described, it is the relation, in perfect proportion, of the whole to its parts. It is a relationship so perfect that its parts are to each other as the whole is to its larger part.

The power of the golden section to create harmony arises from its unique capacity to unite different parts of a whole

Ark of the Covenant

The Divine Proportion is found far back in the stories of the Old Testament. In Exodus 25:10, God commanded Moses to build the Ark of the Covenant:

> *Have them make a chest of acacia wood*
> *two and a half cubits long,*
> *a cubit and a half wide,*
> *and a cubit and a half high.*

These measurements render a shape that is perfectly proportioned according to the Divine Proportion.

OPPOSITE: A woodcut from the book *L'atmosphère météorologie populaire* by Camille Flammarion (1842–1925) shows someone breaking through the medieval world to see the underlying mechanisms that turn the world.

The inscription on this picture reads: "Aristippus the Socratic philosopher being shipwrecked in Rhodes noticed some diagrams drawn on the beach and said to his companions, 'We can hope for the best for I see the signs of men.'"

so that each preserves its own identity, and yet blends into the greater pattern of a single whole.
GYORGY DOCZI, *THE POWER OF LIMITS*

PROBING THE MYSTERIOUS

When humankind first began to probe the mysteries of the universe, they invented new languages, expressing their discoveries and sacred wisdom in art and architecture, songs and chants, and sacred rituals. Their expressions were couched with a profound reverence, for the ability of the heart and mind of human beings to understand the mysteries created an awesome responsibility.

As humans of the past wrestled with questions of who and what we are, they developed several ways to express existential truths. In times much different than our own, when numbers spoke of mystical relationships and divine harmonies, the ratio of Divine Proportion was part of the developing language of mathematics. This was a language of real and symbolic applications, believed to have an extraordinary ability to pry open secrets. As a language, mathematics lies somewhere between the worlds that can be translated to language and those that are described by art—a mysterious, harmonious, almost magical modality in which abstractions become realities and unknowns appear as solutions to practical puzzles.

A practical understanding of relationships and proportions has always had significance for people, mostly because practical applications are so useful in everyday activities. The Divine Proportion is the name given to a mathematical ratio that was originally measured out with rope. It was used to give both the

THE EYE OF HORUS

Cradled in a valley near the mountains of Thebes was a small Egyptian monastery called Ta Set Maat or The Place of Truth. It was built near to the village of Deir el-Medina and housed a community of workmen involved in the construction of the royal tombs.

This wall painting comes from one of the tombs and it shows the udjat-eye, or eye of Horus, holding a lamp from which two flaming streaks emerge. The eye of Horus had mystical significance. Horus was the only son of Isis and Osiris; he swore to avenge his father's death at the hands of Osiris' brother Seth. During a ferocious battle Seth ripped out Horus' eye, tore it into six pieces and scattered them across Egypt. Horus returned the compliment by castrating Seth. The gods finally intervened and designated Horus as King of Egypt. Then they instructed Thoth, the god of learning and magic, to reassemble Horus' eye. Thus it became a symbol of wholeness, clear vision, abundance, and fertility.

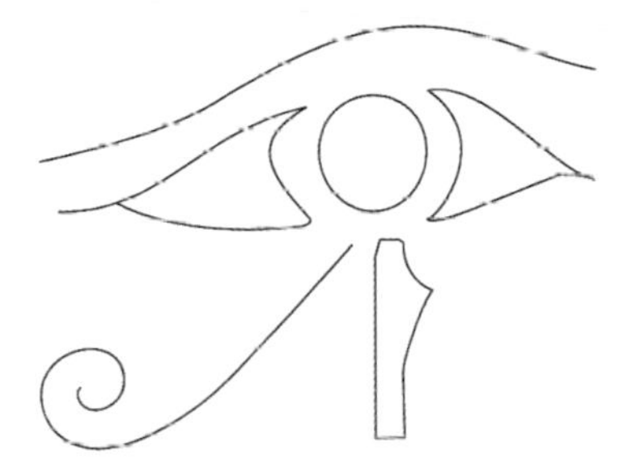

This hieroglyph was used by scribes to symbolize fractions in their measurements.

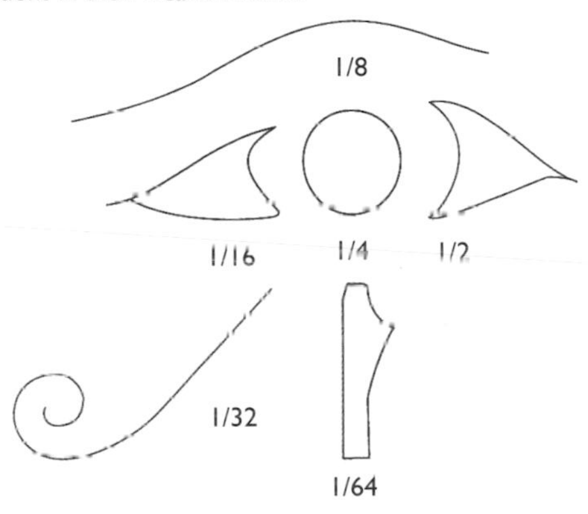

Each element represented a fraction from 1/2 to 1/64 and could be used in any number of combinations. A story is told that one day an apprentice scribe told his master that the total fractions of the eye of Horus did not add up to one, but rather to 63/64. The master replied that Thoth would make up the missing 1/64 to any scribe who sought and accepted his protection.

These six parts also correspond to the six senses of touch, taste, hearing, thought, sight, and smell.

Orpheus was a mythical figure and one of the chief poets and musicians of ancient Greece. He was the inventor of the lyre, and his voice was so sweet that by his singing he was able to charm the wild beasts, to persuade the trees and rocks to move from their places, and to stop the rivers in their course. Music, in a way that does not lend itself to geometric analysis, has always been associated with the laws of proportion.

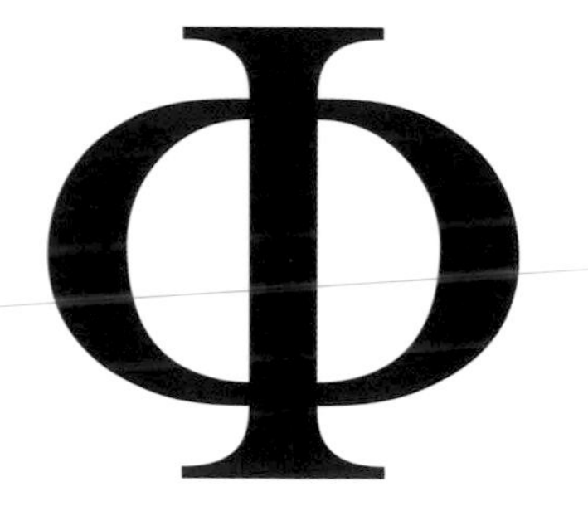

The Greek symbol Φ is the equivalent of the English sound "ph". It sounds like fly without the "l" and is used to represent the proportion that Euclid described in Chapter 5 of *Elements*. This proportion is called the Divine Proportion. It is also known as the Golden Proportion, the Golden Mean and the Golden Ratio.

Great Pyramid at Giza and the Parthenon in Athens their very harmonious shapes.

Although this proportion appears to have been understood for centuries, it was first articulated mathematically by Euclid of Alexandria (ca. 325–265 B.C.E.) in his book, *Elements*. In the fifth chapter, Euclid drew a line and divided it into what he called its "extreme and mean ratio":

> *A straight line is said to have been cut in extreme and mean ratio when, as the whole line is to the greater segments, so is the greater to the lesser.*

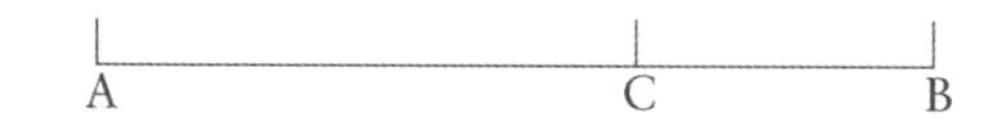

In other words AB / AC = AC / CB. When we speak of this equation we say that AB is to AC as AC is to CB.

This proportion, articulated by Euclid, tells us that the ratio of the whole line AB to its larger part AC is the same as the ratio of the larger part AC to the smaller part CB. When calculated mathematically this gives a ratio of approximately 1.61803 to 1, which can also be expressed as $\frac{(1+\sqrt{5})}{2}$ or Φ.

Divine Proportion in Design

USING THE RELATIONSHIP OF EUCLID'S LINE we can draw a rectangle in which one side is 1 and the other is Φ,

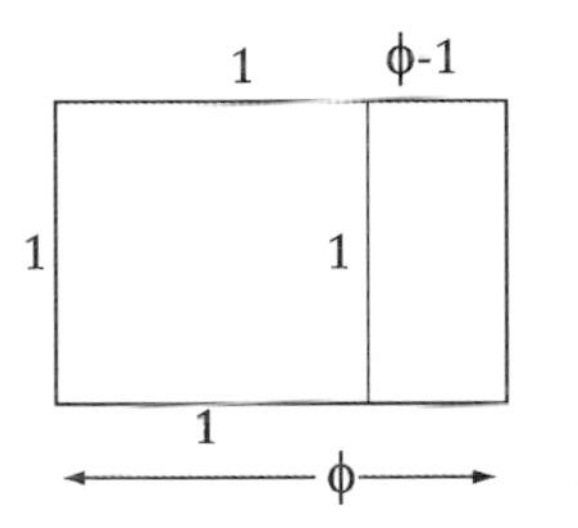

and a pentagram in which the segments a, b, c, d in order of decreasing lengths are in a ratio of 1.618… or Φ.

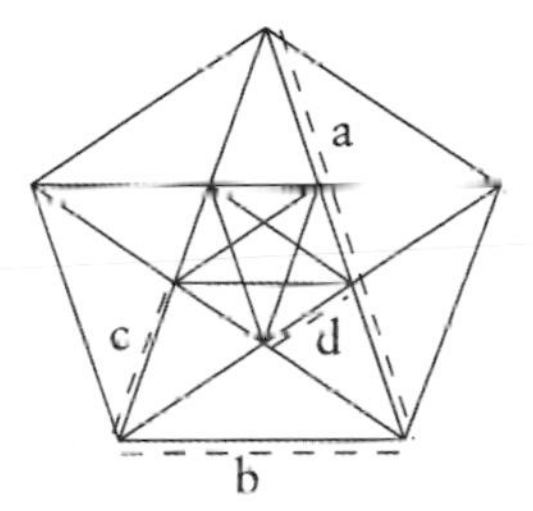

We can also construct the Golden Spiral that emerges when the rectangles or triangles are nested together:

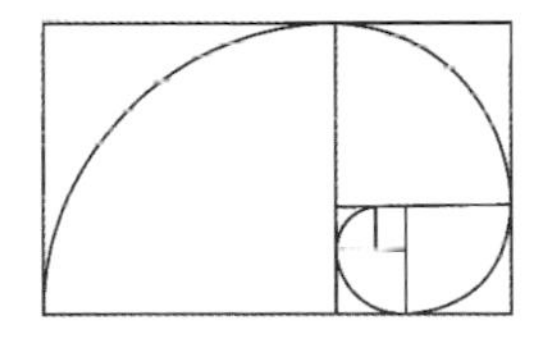

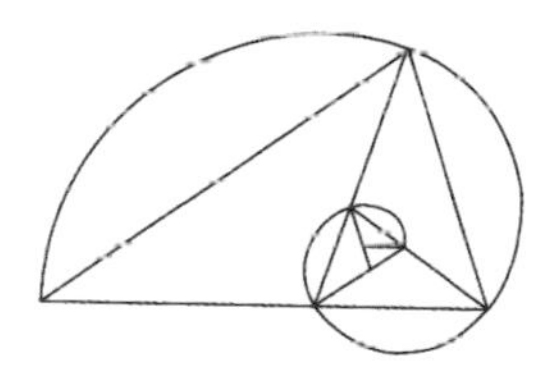

Rectangles based on Φ are embedded in the Parthenon's design.

The pentagram whose segments are related to Φ can be seen in the center of an apple.

The Golden Spiral that is based on Φ describes the way in which many shells and the horns of some animals are formed.

EUCLID OF ALEXANDRIA

EUCLID OF ALEXANDRIA is the leading mathematics teacher of all time. His book, *Elements*, is still used to teach geometry. Little is known of Euclid's life except that he taught in Alexandria, Egypt. Proclus (410/412–485 C.E.), the last major Greek philosopher, said of Euclid:

This man lived in the time of the first Ptolemy; for Archimedes, who followed closely upon the first Ptolemy makes mention of Euclid, and further they say that Ptolemy once asked him if there were a shortened way to study geometry than the "Elements," to which he replied that there was no royal road to geometry.

Proclus tells us that Euclid was younger than Plato and was

... in sympathy with this philosophy, whence he made the end of the whole "Elements" the construction of the so-called Platonic solids.

We will discover more about the Platonic Solids later. For many centuries they were thought to be a perfect description of the universe. Quite naturally they included the Divine Proportion.

Stobaeus (a Greek anthologist from the second half of the 5th century C.E.) gives us one more piece of information about Euclid:

Someone who had begun to learn geometry with Euclid, when he had learnt the first theorem, asked Euclid "What shall I get by learning these things?" Euclid called his slave and said "Give him threepence since he must make gain out of what he learns."

Though there never were a circle or triangle in nature, the truths demonstrated by Euclid would for ever retain their certainty and evidence.

DAVID HUME (1711–1776),
SCOTTISH PHILOSOPHER AND HISTORIAN

Euclid's ability to couch logic in simple language to express laws that had never been defined before was a significant historical development. His book, *Elements*, is one of the most important works of science in the history of humankind, for it marks the beginning of a new way of thinking—one based on empirical thought. Since Euclid's time, generations of successors have continued the effort to unveil the mysteries of the universe and to describe and confirm these mysteries through numbers. Alongside the wonderful history of these discoveries lies the story of the Divine Proportion.

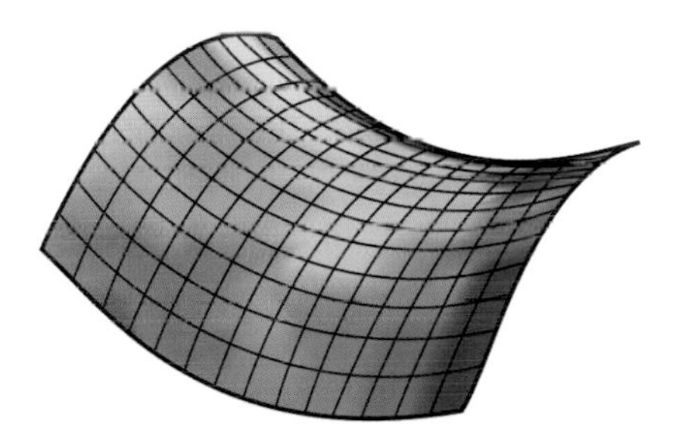

NON-EUCLIDEAN GEOMETRIES

The difference between Euclidean and other geometries is their treatment of parallel lines. In Euclidean geometry, parallel lines never meet and they are always separated by the same distance. In the geometries of different shapes there are instances where there are no parallel lines and two straight lines will always meet somewhere, and there are instances in which parallel lines can exist, but they will not always be the same distance from each other.

STUDYING THE UNIVERSE

Euclid's book, *Elements*, begins with definitions and five postulates or axioms. The first three axioms refer to construction, stating, for example, that it is possible to draw a straight line between any two points. These axioms implicitly assume the existence of points, lines, and circles. The existence of other geometric shapes is deduced from the fact that these points, lines, and circles exist.

The famous fifth, or parallel, axiom states that one and only one line can be drawn through a point so that it is parallel to a given line. This axiom was challenged in the 19th century, and non-Euclidean geometries (based on the fact that there are curves in existence) began to be studied.

After Euclid and Plato came a long line of people who studied the Divine Proportion. These included Kepler, Luca Pacioli, Roger Penrose and the never-to-be-forgotten Leonardo of Pisa, now known as Fibonacci. In the 21st century the most stunning and relevant development in the study of the Divine Proportion

There are many approaches to learning. Some people inquire through scientific methods, others through inquiry into the nature of who they are.

Scholars study from books.

A dervish contemplates his interiority.

lies in its connection to the sophisticated world where astrophysics has met the metaphysical. Once again mathematics has prodded into the mysterious, and our worldview is being confronted by new discoveries that challenge old understandings. In a world that for twenty-five hundred years has developed an extraordinary compendium of knowledge based on principles of logic and rational thought, we now find ourselves faced by the realization of physicists that experience, not knowledge, is the real key to discovering universal principles. An old question asks, if there is a flower that no one sees, how do we know it exists? We have no way of proving for certain that it does. By the same token, we do not know for sure that any given mathematical law is absolutely true in every case, because there are an infinite number of applications.

As we begin to understand this, and to peer with new eyes and listen with different ears, we discover that while we presume we are living in the same universe that existed long before Euclid began to describe it mathematically, we cannot be certain about that. The only universe we can really describe is the one we are experiencing now.

There is an ancient Islamic saying that echoes these new mathematical ideas. It says that Allah created the universe so He could study Himself.

The reason this changing viewpoint is so important to us now is that as we begin to study the Divine Proportion, we must remember that one of its most interesting aspects is the relation it speaks of when we put ourselves and the universe into the equation.

How to Divide a Line and Find Divine Proportion

To divide any line at the point that divides it into its mean and extreme ratio let AB be the given straight line. Draw BD perpendicular to AB so that $BD = \frac{AB}{2}$. Join AD.

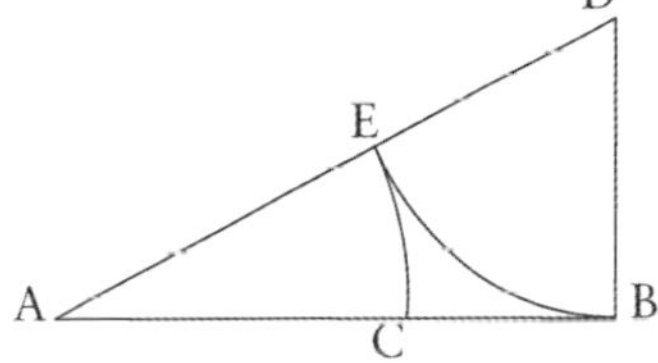

With center D, and radius DB, draw an arc cutting DA at E. With center A, and radius AE, draw an arc cutting AB at C. The point where C intersects the line AB divides this line into its mean and extreme ratio.

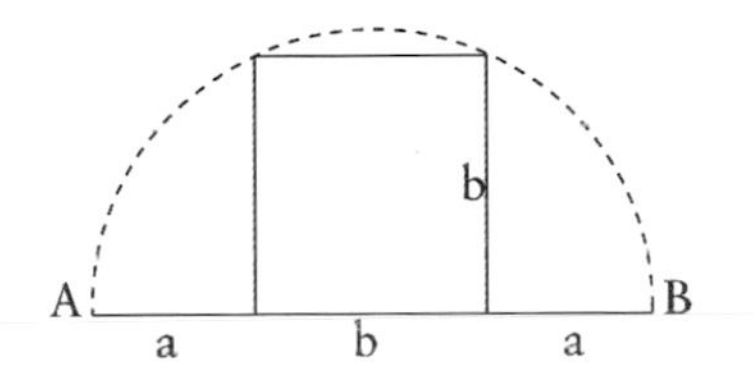

The proof of this can be worked out if a square is drawn on AB. The center of this segment becomes the center of a semicircle.

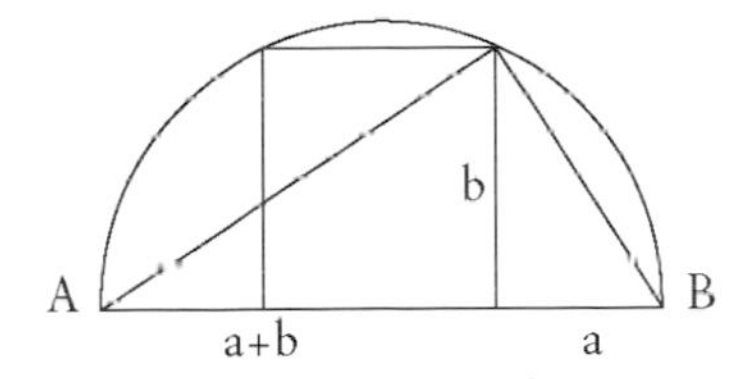

The triangles that can be drawn from the point where line b meets the circumference are called similar triangles. One of these has sides (a+b) and b, the other has sides b and a. These two triangles have the same angles and their sides are all proportional. They are known as Golden Triangles. The line AB is proportional to the segment (a+b) as (a+b) is to the segment a.

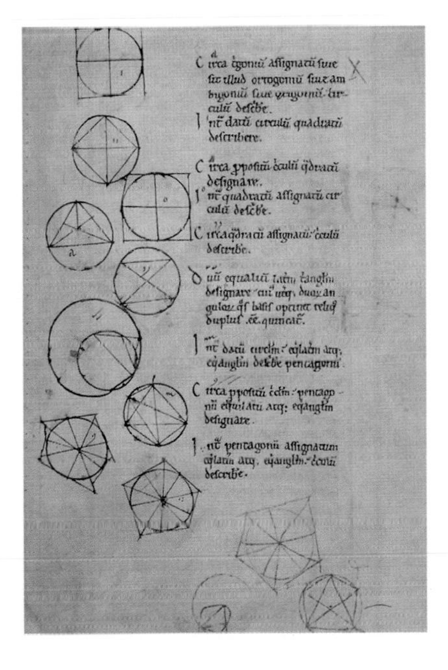

A page from a medieval Latin copy of Euclid's *Elements* in Arabic

TIMELINE OF MANY PEOPLE WHO HAVE MADE IMPORTANT

PHIDIAS
490–430 B.C.E.

Phidias was a Greek sculptor as well as a mathematician who helped direct the building of the Parthenon. He is said to have applied Divine Proportion to the design of the sculptures found there.

PLATO
(427–347 B.C.E.)

In Timaeus, Plato described five possible regular solids as the basis for the harmonious structure of the universe. The Divine Proportion plays a crucial role in the dimensions and formation of some of these solids.

EUCLID
(325–265 B.C.E.)

In *Elements*, Euclid gave the first recorded definition of the Divine Proportion.

FIBONACCI
(1170–1250)

Fibonacci discovered the numerical series that is now named after him. This series is closely connected to the Divine Proportion.

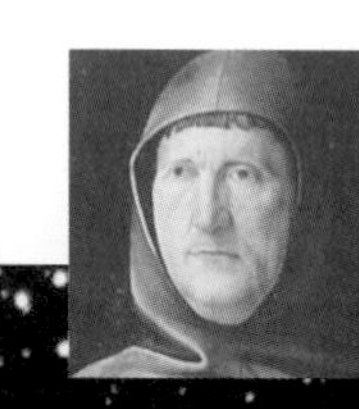

LUCA PACIOLI
(1445–1517)

In *Divina proportione,* Pacioli gives reasons why the Golden Ratio should be called the Divine Proportion.

KEPLER
(1571–1630)

Kepler says of the Divine Proportion that it is a "precious jewel."

DISCOVERIES CONCERNING THE DIVINE PROPORTION

CHARLES BONNET (1720–1793)

Bonnet pointed out that in the spiral phyllotaxis of plants the number of spirals going clockwise and counter clockwise were frequently two successive Fibonacci numbers.

MARTIN OHM (early 19th century)

Ohm is believed by many to have been the first to formally use the words "golden section" to describe the Divine Proportion.

EDOUARD LUCAS (1842–1891)

Lucas officially dubbed the numerical sequence now known as the Fibonacci sequence with its present name.

MARK BARR (20th century)

The first Greek letter in the name of Phidias, Φ (Phi), was given to the ratio of Divine Proportion by Mark Barr.

ROGER PENROSE (b. 1931)

In the field of periodic tilings Penrose found a symmetry that uses the Divine Proportion that led to a new discovery about quasicrystals.

Highlights in the Story of Divine Proportion

The following is a short description of the highlights that are most significant in the story of the Divine Proportion, and which will be told in much greater detail later in this book.

The miracle is that the universe created a part of itself to study the rest of it, and that this part in studying itself finds the rest of the universe in its own natural inner realities.
John C. Lilly (1915–2001), scientist

While the proportion now known as the Divine Proportion has always existed in mathematics and in the workings of nature, we do not know exactly when it was first discovered and applied by humankind. It is reasonable to assume that some or many of its aspects have been discovered, lost, and rediscovered many times throughout history, which helps to explain why it goes under different names. Certainly we will see when we look at the life of Fibonacci that tremendous amounts of information the Greeks had compiled was lost to the whole of Europe during the Middle Ages. It was only due to Fibonacci's brilliant and curious nature that he put lost pieces back together in his travels and came up with a brilliant new discovery.

The writings of HERODOTUS (5th century B.C.E.), the world's first historian, have led many to believe that the Egyptians used the Divine Proportion in the design of the Great Pyramid and that the Greeks, who called it the Golden Section, based the entire design of the Parthenon upon it.

PHIDIAS (490–430 B.C.E.), a Greek sculptor and mathematician, is said to have studied the Golden Section and applied it to the design of his sculptures for the Parthenon. Although everything he created has disappeared, we know of him from others who wrote about his work. His most famous piece—the statue of Zeus at Olympia—was one of the Seven Wonders of the Ancient World.

PLATO (ca. 427–347 B.C.E.), in his views on natural science and cosmology, informed us of a certain proportion (now known as the Golden Mean) that is the most binding of all mathematical relationships and the key to the physics of the cosmos.

EUCLID (325–265 B.C.E.), in *Elements*, wrote his theorem about dividing a line in the extreme and mean ratio. This theorem, along with its proof, was the first time the Divine Proportion had been put into mathematical language. Euclid also linked this ratio to the construction of a pentagram.

FIBONACCI (1170–1250), an Italian originally named Leonardo of Pisa, discovered the unusual properties of a numerical series (0, 1, 1, 2, 3, 5, 8, 13, 21...) that now bears his name, though it is not certain that he realized its connection to the Divine Proportion. His book *Liber abaci* played a pivotal role in the European adoption of the Arabic decimal system of counting over the system of Roman numerals that had been used before.

Leonardo of Pisa, now known as Fibonacci, published a book in the 13th century with a fictitious problem about the breeding of rabbits. The solution to the problem is a series of numbers called the Fibonacci sequence, 0, 1, 1, 2, 3, 5, 8, 13.... These numbers are intricately related to the Divine Proportion.

LUCA PACIOLI (1445–1517), geometer and friend of the great Renaissance painters, "rediscovered" the "golden secret" and proposed, in his book *Divina proportione*, to call it the Divine Proportion. Leonardo da Vinci provided illustrations for the work, with beautiful drawings of the five Platonic Solids. It was probably Leonardo who first called this proportion the *sectio aurea*, which is Latin for golden section.

Many Renaissance artists used the Divine Proportion in their paintings and sculptures to achieve balance and beauty. According to several accounts, Leonardo da Vinci used it to define all the fundamental proportions of *The Last Supper* and *Mona Lisa*.

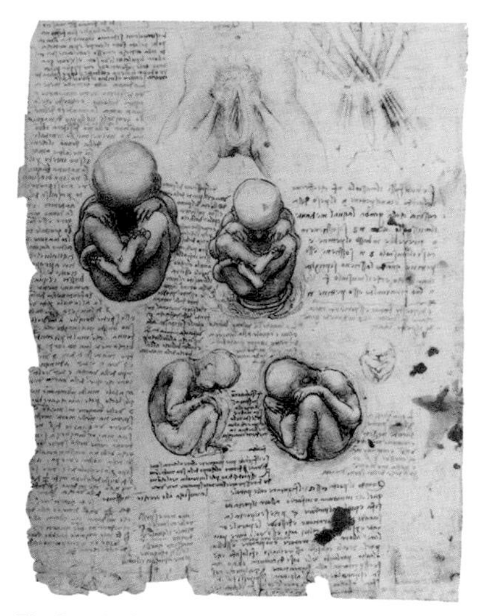

The Fetus by Leonardo da Vinci

The probes of artists, scientists, mathematicians, and philosophers have uncovered many of nature's secrets. They have discovered laws of natural balance and properties of regeneration that are inherent in the Divine Proportion.

The Side Show by Georges Seurat (1859–1891)

Seurat makes obvious use of the Divine Proportion in the layout of his canvas.

Artists both ancient and modern have used the Divine Proportion to design their art. They have incorporated its symbols into the messages they wish to express and have used its underlying harmony to achieve symmetry and balance. Observing its principles in the natural world they have attempted to express the timeless beauty of the Divine Proportion as it expresses itself in nature's dynamic principles.

JOHANNES KEPLER (1571–1630), discoverer of the elliptical nature of the orbits of the planets around the sun, uncovered the connection between the Fibonacci sequence and the Divine Proportion, showing that the ratios of successive terms of the Fibonacci sequence tend to approach the Golden Ratio. He also looked at plants and incorrectly concluded that the Fibonacci sequence propagated itself in their seeding capacity but correctly surmised the intrinsic involvement of the sequence in plant growth.

CHARLES BONNET (1720–1793) described the arrangements of leaves on plants in his *Recherches sur l'Usage des Feuilles dans les Plantes* and pointed out that the number of spirals going clockwise and counterclockwise in the phyllotaxis of plants were frequently two successive Fibonacci numbers.

MARTIN OHM (late 18th to 19th centuries), a German mathematician, and brother of George Ohm (the physicist for whom the unit of electrical resistance, the ohm, was named), is believed to have been the first to formally use the words "Golden Section" to describe this wonderful proportion—in a footnote in the 1835 edition of his book *Die reine Elementar-Matematik*:

EXPERIMENTS OF GUSTAV FECHNER

Gustav Theodor Fechner (1801–1887) was a pioneer of experimental psychology. Curious about the Divine Proportion, he performed many experiments in his investigation of its appeal. Beginning by taking the measurements of thousands of rectangular objects, such as books, boxes, and buildings, he found that the average rectangle used was close to the ratio of Φ. He later found, by showing a series of rectangles to a number of people, that most people preferred the same one.

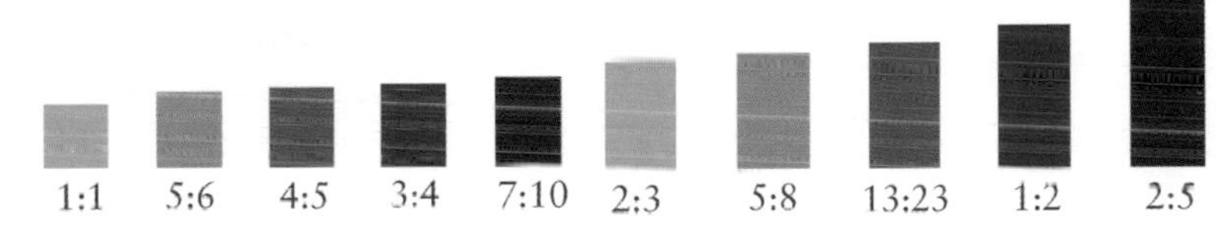

Fechner conducted many experiments such as this one in which he asked people to point to the rectangle that they found to be most pleasing. Readers can do this experiment and compare choices to the graph below which charts the results of Fechner's findings.

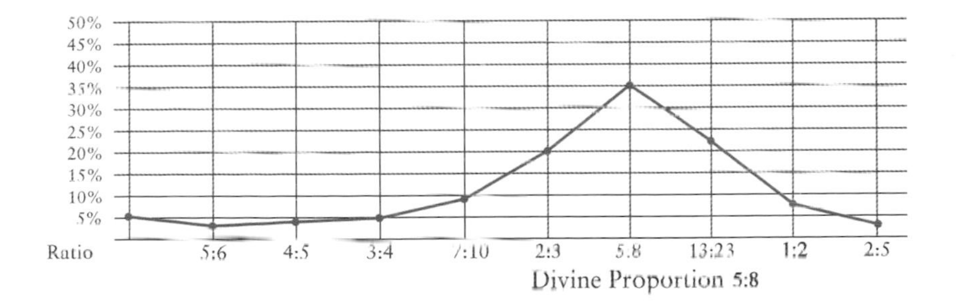

Jay Hambidge (1867–1924), an art theorist, pioneered the technique of searching for certain typically "sacred" geometric ratios among the measurements of ancient artifacts. Like all sacred geometry detectives, he had to perform endless calculations on dozens of items of differing sizes, establishing beyond doubt that it was ratio, not measurement, that determined the relative dimensions. The result of his work was the definition of a principle called "dynamic symmetry" that incorporates the Divine Proportion and leads, Hambidge said, to organic symmetry in art.

> *One is also in the habit of calling this division of an arbitrary line in two such parts the golden section; one sometimes also says in this case: the line r is divided in continuous proportion.*

It is fairly clear from Ohm's footnote that the term "golden section" does not come from him. However, it seems that 1835 marks the first appearance of the term in public usage. An earlier edition of the same book did not use the phrase.

EDOUARD LUCAS (1842–1891), French mathematician for whom the Lucas Series is named, officially gave the Fibonacci sequence its name.

MARK BARR used the Greek letter Φ (Phi) to designate the Divine Proportion at the beginning of the 20th century. At that time this ubiquitous proportion was known as the Golden Mean, Golden Section, and Golden Ratio, as well as the Divine Proportion. Phi is the first letter of the name of Phidias, who is thought to have used the golden ratio in his sculptures. It is also the Greek equivalent to the letter "F," the first letter of Fibonacci.

ROGER PENROSE (b. 1931), an English mathematical physicist, conjectured the cosmic censorship hypothesis, which suggests that the universe protects us from the inherent unpredictability of things like black holes by hiding them from our view. In the field of tiling, which describes methods of achieving symmetrical tiling patterns, Penrose is known for his discovery of "Penrose Tiles." This discovery allowed surfaces to be tiled in a type of symmetry that previously had been thought to be impossible.

This discovery led to a breakthrough understanding about matter and an ability to describe what have been termed quasicrystals—significantly all tied into the Divine Proportion.

This short list is not complete without the mention of some correspondence between architect Le Corbusier and scientist Albert Einstein. Le Corbusier had proposed that what industry needs is a system of proportional measurement that would reconcile the needs of the human body with the beauty of the Divine Proportion and he designed a concept called Modular. After listening to Le Corbusier's description of the concept, Einstein wrote a note saying that it is a "scale of proportions which makes the bad difficult and the good easy."

> *All things are full of signs, and it is a wise man who can learn about one thing from another.*
>
> PLOTINUS (205–270), ROMAN PHILOSOPHER

Quasicrystal World by contemporary artist Matjuska Teja Krasek

> *A proportion that gives us an experience of the aesthetic appeal is the Golden Mean. I believe that one of the reasons we perceive it as beautiful may lay in the fact that it appears in the proportions of our own bodies, and that we can find it in the world that surrounds us. Therefore, it is not surprising that it was incorporated in artworks throughout history; and it should not be ignored now at the end of the twentieth century.*
>
> MATJUSKA TEJA KRASEK, SLOVENIAN ARTIST

Chapter 2

Pythagoras and the Mystery of Numbers

There is geometry in the humming of the strings...
there is music in the spacing of the spheres.
PYTHAGORAS

NUMBERS SERVE TWO PURPOSES. First, their use is practical. They comprise the tools we use to count and measure. Second, they are the means by which many people have attempted to understand the mysterious and the unexplainable.

In prehistoric times, as the connection between the moon's phases and all the growing cycles of life was observed, a language of symbols emerged that could represent and keep track of natural phenomena. As needs grew and demanded some sort of record keeping, the language grew more complex, and the number sense developed. Concepts of one, two, three, or four were relatively easy to express, but beyond that a specific ability to count was required.

Studies on number perception in a wide range of beings from crows to humans have found that human adults are normally able to count from one to four with no training at all. Beyond that we

Creation of the World and Expulsion from Paradise by Giovanni di Paolo (1400–1482)

Di Paolo shows a cosmos that is divided into several spheres or heavens—the sun, moon, and planets, the fixed stars, the *Primum mobile* that regulated the motion of all the spheres beneath it, and the Empyrean heaven, the home of God and the angels. The number of spheres to identify the heavens in pictures of this sort vary because theologians could never decide whether the Empyrean occupied a sphere, or whether it was infinite and unknowable. This created quite a big problem for artists. Di Paolo shows no sphere for the Empyrean, just a region beyond the last ring, implying it is infinite.

OPPOSITE: A scene from the Satirical Papyrus (ca. 1295–1069 B.C.E.) probably from Thebes, showing a lion and an antelope playing senet, the best-known board game of ancient Egypt.

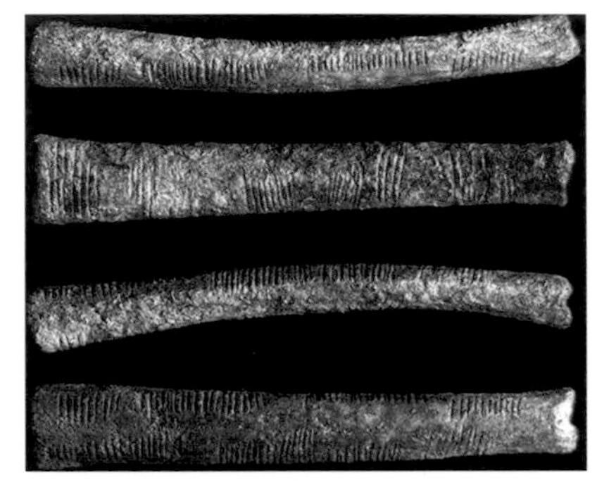

The history of how we count and what numbers mean to us goes back into the misty origins of human life and civilization. The Ishango bone is the oldest known object containing clear numerical carvings. It was discovered in the Congo and is 20,000 years old.

A clay tablet from the Old Babylonian Period (ca. 1900–1600 B.C.E.) has a mathematical table with four columns and fifteen rows engraved on it using cuneiform script. Clay made a perfect tablet for it was plentiful and as long as it remained moist, calculations could be erased. Once it had hardened, the tablet was either thrown away or used as a building material.

have to learn. This is a two-step process. First, we develop a system of counting and the skills of manipulating numbers in that system. In order to memorize and communicate information gleaned from this, we then develop a way of linguistically naming the individual items. Once a system is in place (and learned) and the items are named (and learned), we devise a scheme for writing the numbers down, making it much easier to manipulate them.

It is not necessary to count in the way that we do now in order to make sure all the sheep have returned home or keep track of passing time. Even without words and the abstract concepts of numbers, all sorts of effective techniques have been devised to facilitate counting. Groups of people in present-day America, Asia, and Africa use languages with very few word-numbers to manage their affairs. With a language limited to simple concepts of "one," "two," and "many," they use notches on bones or sticks, draw lines, use piles of pebbles or shells, or tick off parts of the body (fingers, toes, elbows, eyes, nose) to keep track of specifics.

Of all the devices used to record calculations, the earliest discoveries are clay tablets from areas in present-day Iran and Iraq (ca. 3200 B.C.E.), where unbaked clay was marked with shapes assigned to numerical values. The tablets found in Iran were based on a counting system of a base of ten, while those found in the nearby area in Iraq used a base system of sixty. Both of these systems are still used by us today—base ten in our use of a decimal system that counts things by tens and base sixty in our use of sixty minutes to tell time and to measure degrees in a circle.

We know from remains that the Babylonians were highly proficient in algebra, geometry, and astronomy. They even reckoned with

THE NUMBER SENSE

EXPERIMENTS BY specialists in animal behavior have shown that some animals appear to have an ability to perceive quantities. This is sometimes referred to as the number sense, and it allows an animal to determine the differences in size between two small collections of similar objects. It also allows them to determine that a collection is not the same after some objects are removed. Both domestic and wild animals have shown that mothers can definitely determine when one of their young is missing from the group.

Birds can be trained to determine the number of seeds in different piles up to the number five. There is a well-known story of a crow—a bird well recognized for intelligence and cunning—that illustrates this ability.

A crow had built a nest in the watchtower of a young squire's estate. The squire wanted to remove the nest, but each time he approached the tower the crow would leave, returning once the squire had left. The squire finally came up with a plan to fool the bird. He took a man into the tower with him and then asked the man to leave—but the crow would not return to the tower until both men had gone. The squire repeated his experiment during the following days with two, three, and even four men and still the crow would not be fooled. Finally five men went into the tower. One by one they left and at the point when four had gone—and one still remained—the crow lost count and returned to its nest, and the squire rid the tower of the crow.

This chart shows how we perceive

numbers. Looking at it we easily notice which groups of items are recognizable as a numerical quantity and which are not. Some groupings are clear to our number sense, and some must be counted.

EARLY CALENDARS

KEEPING TRACK OF THE MOON was a major concern of Neolithic peoples, for knowing its cycles was a key to predicting the seasonal changes. Finding ways to keep track of astronomical observations has probably been the single biggest influence on the development of mathematics.

Five thousand years ago, Sumerians in the Tigris-Euphrates valley had a calendar system that divided the year into 30-day months, the day into 12 periods (each corresponding to 2 of our hours), and these periods into 30 parts (each like 4 of our minutes).

We have no written records of Stonehenge, built over 4,000 years ago in England, but its alignments show its purposes apparently included the determination of seasonal or celestial events, such as lunar eclipses, and solstices.

The earliest portion of the Stonehenge complex dates to approximately 2950–2900 B.C.E. Postholes dated from about 2900 to 2400 B.C.E. indicate there were timber structures in the center of the monument and at the northeastern entrance, but no clear patterns or configurations are discernible that would suggest their shape, form, or function. In the last phase of its construction (ca. 2550–1600 B.C.E.) huge stones were set, and reset, after being transported a distance of 50 miles.

Neo-Assyrian calendar of good and evil days (705–681 B.C.E.)

The earliest Egyptian calendar was based on the moon's cycles, but around 3100 B.C.E. they realized that the "Dog Star" (which we call Sirius) rose next to the sun every 365 days, about the same time as the annual inundation of the Nile began. With this understanding they devised a 365-day calendar.

Papyrus calendar of good and evil days from 1200 B.C.E.

irrational numbers and infinite decimal expansions. Their questions and methods of solution were stated rhetorically—in words rather than symbols and equations.

Egyptian Civilization and Mathematics

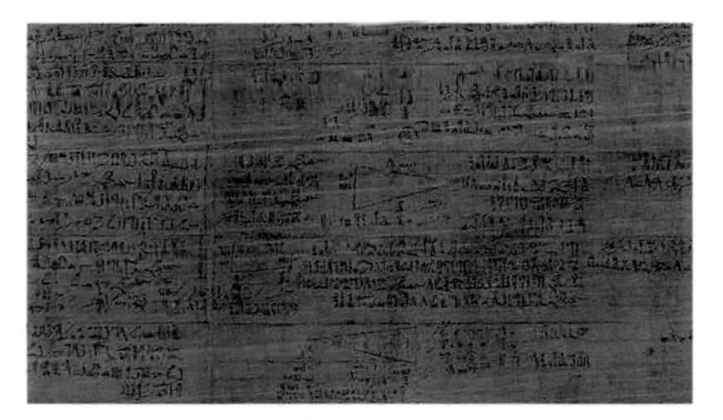

The Rhind Papyrus dates back to 1650 B.C.E. and describes methods and problems used by the Egyptians.

The Egyptians had no currency so transactions were carried out by trading goods—most commonly beer and bread. One problem in the Rhind Papyrus describes how to divide nine loaves of bread among ten people. Whereas we might calculate that each man should receive 9/10 of a loaf and thus cut one tenth from each loaf and give all the small bits to one man, the Egyptians had a different solution. The solution in the papyrus says that 9/10 = 2/3+1/5+1/30, requiring more cuts to be made to each loaf but resulting in each man receiving identical sized pieces.

Egypt, a civilization that spanned about 4,000 years, left little written evidence of its mathematics, for the papyrus they used to write on was very fragile. However, the Rhind Papyrus from 1650 B.C.E. gives a very clear indication of the extent of their calculations and the methods they used. The Rhind Papyrus is a scroll almost twenty feet long and a foot wide. It is named after Alexander Henry Rhind, who purchased it in Egypt in 1858, and is sometimes called the Ahmes Papyrus, in honor of the Egyptian scribe who allegedly copied it from another document two centuries older. The opening sentence claims that the text is "a thorough study of all things, insight into all that exists, knowledge of all obscure secrets." The papyrus contains eighty-seven problems and their solutions and is written in everyday script.

A remarkable amount of practical mathematics, some of it extraordinarily sophisticated, was developed by the early agricultural civilizations of Egypt and Mesopotamia and countries farther east. The ways in which these people understood their mathematics was always tightly intertwined with developments in religious and philosophic thought. For the development of our own systems, we look back to the mathematical foundations developed by the ancient Hellenic Greeks. They, in turn, traced all human inventions—from calculus, geometry, astronomy, and dice games to writing—back to the Egyptians. The priests of Egypt began to keep written records between 4000 and 3000 B.C.E. More than two thousand years later, the poems of Homer

SCHOOLS OF LEARNING

Luxor is today's name for the ancient city of Waset that was partially built on the site of ancient Thebes. The Temple of Waset can be described as the world's first university. Egypt's pharaohs and the high priests went there to be trained in the "Mystery Systems."

The city of Alexandria was founded by Alexander the Great. His successor, Ptolomy II Soter, opened the Museum or Royal Library of Alexandria in 283 B.C.E. It was destroyed in the civil war that occurred at the end of the 3rd century C.E. under the Roman emperor Aurelian.

The Museum was modeled after Aristotle's Lyceum in Athens and included lecture areas, gardens, a zoo, and shrines for each of the nine muses. Some estimate that at one time the Library of Alexandria held over half a million documents from Assyria, Greece, Persia, Egypt, India, and many other nations. Over one hundred scholars lived at the Museum to research, write, lecture, or translate and copy documents.

were still being circulated through an oral tradition, for the Greeks still had not developed a method of writing.

Herodotus, the Greek researcher and storyteller, tells us in his *History* that:

> *The king moreover (so they said) divided the country among all the Egyptians by giving each an equal square parcel of land, and made this his source of revenue, appointing the payment of a yearly tax. And any man who was robbed by the river of a part of his land would come to the sesostris* [the pharaoh] *and declare what had befallen him; then the king would send men to look into it and measure the space by which the land was diminished, so that thereafter it should pay in proportion to the tax originally imposed. From this, to my thinking, the Greeks learned the art of geometry.*

Proclus reports in his *Commentaries on Euclid's Elements* that Eudemus, a disciple of Aristotle, said, "We shall say, following the general tradition, that the Egyptians were the first to have invented Geometry, (that) Thales, the first Greek to have been in Egypt, brought this theory thereof to Greece."

The importance of Egypt as a source of wisdom cannot be underestimated. The Temple of Waset (present-day Luxor), the world's first university, known as "the septer," was built during the reign of Amenhotep III (about 1391 B.C.E.). At its zenith it educated 80,000 students. Thales of Miletus, Plato, Aristotle, Socrates, Euclid, Pythagoras, Hypocrates, Archimides, and Euripides all studied there. The Greek St. Clement of Alexandria once said that if you were to write a book of 1,000 pages, you would not be able to put down the names of all the Greeks who went to Kemet. (Kemet is the name the Egyptians gave their own country; the Greeks called it Egypt.)

Early Greek Travelers to Egypt

Herodotus is the Greek author known to us as the world's first historian. His great narrative *History* has as its subject the wars between Greece and Persia (499–479 B.C.E.) and their preliminaries. As it has survived, the *History* is divided into nine books that describe the background and history of the wars.

We know little about Herodotus apart from what he wrote, but he was obviously well educated and traveled extensively. It is believed he was born in Halicarnassus, a Greek city in southwest Asia Minor that was then under Persian rule. As a reason for writing his *History* it is thought that as a young man he was deeply impressed not only by the great size of the Persian Empire but also by the nature of its army, which was united in a single command, in contrast to the Greek forces with their many political divisions. In an attempt to explain this difference to his readers, Herodotus weaves a complex history that includes fairy tales, gossip, and legends in a description of the empire and the many peoples the Persians had conquered.

Thales of Miletus is one of the legendary Seven Wise Men, or Sophoi, of antiquity—a group of ancient Greek sages of the 7th and 6th centuries B.C.E. He is primarily remembered for developing a cosmology based on water as the essence of all matter, with the earth a flat disc floating on a vast sea.

There are no writings of Thales that have survived and no contemporary sources exist, thus, his achievements are difficult to assess. Numerous sayings are attributed to him, such as "Know thyself" and "Nothing in excess."

According to Herodotus, Thales was a practical man who used his knowledge of geometry to measure the Egyptian pyramids and to calculate the distance of ships from shore. It is also said that he predicted a solar eclipse on May 28, 585 B.C.E.

Thales has been credited with the discovery of five geometry theorems: (1) that a circle is bisected by its diameter, (2) that angles in a triangle opposite two sides of equal length are equal, (3) that opposite angles formed by intersecting straight lines are equal, (4) that the angle inscribed in a semicircle is a right angle, and (5) that a triangle is determined if its base and the two angles at the base are given.

Again, these achievements are difficult to assess because of the ancient practice of crediting particular discoveries to men with a general reputation for wisdom.

Aristotle claimed that Thales was the founder of philosophy, based on the development of Thales' cosmology. From a modern perspective, the significance of Thales' views lies less in his cosmology than in his attempt to explain nature by means of physical phenomena and in his search for causes within nature itself rather than in the caprices of the gods. In some part, Thales' importance as a philosopher lies in his bridging the worlds of myth and reason.

Development of Athens

In the 8th century B.C.E., a group of noblemen became very wealthy and overthrew their king. It wasn't long before rich farmers were making money, but average farmers had fallen deeply into debt. To pay their debts they sold their children, their wives, and even themselves into slavery, both in Athens and abroad.

Recognizing the danger of the situation, the people of Athens agreed to hand over all political power to a single individual, Solon, in 594 B.C.E. He dismissed all outstanding debts and freed the Athenians. Although he brilliantly reformed the government, he could not solve the economic crisis, and within a few years Athens collapsed into anarchy. Peisistratus swept into power and his tyranny was as important to the foundation of Athenian democracy as Solon's reforms had been. He was both a military leader and a man devoted to cultural reform. He sought out poets and artists in order to make Athens a cultural society. After many years and more reform—by 500 B.C.E.—Athens had pretty much established its culture and political structure and was more or less a democracy. It had become a rich and beautiful city, the center of art and literature, and for the next one hundred years the ancient world was politically and culturally dominated by Athens.

The Classical Period of Ancient Greece

The end of the Mycenaean period (ca. 1600–1100 B.C.E.) saw the end of monarchy and the emergence of small city-states in ancient Greece. In some of these city-states, a restricted number of men exercised political power (thus creating a political system called an oligarchy, meaning "rule by the few"). Some were ruled by men who seized power and others created early forms of democracy ("rule by the people") and gave all male citizens the power to participate in governing. The people of Athens established Greece's most famous democracy in which the individual freedom of citizens flourished to a degree unprecedented in the ancient world.

The emergence of this new form of governance brought a period of political, philosophical, artistic, and scientific achievements that formed the legacy upon which Western civilization is based. It lasted from about 480 B.C.E., the time of the defeat of the Persian invaders, until the death of Alexander the Great in 323 B.C.E.

One question that dominated early Greek philosophy was a dispute as to which is more basic, arithmetic (which was based on numbers) or geometry (in which concepts can be described as shapes). Underlying this dispute was a question that asked: Is the universe made up of discrete items that can be counted or does it consist of continuous substances that can only be measured? This dichotomy was presumably inspired by a linguistic distinction, analogous to one we have in English, between nouns such as "apple," that can be separated into single units, and nouns such as "water," that indicate a mass. In order to fully understand how this question came about and the enormous relevance of the mystical concept of numbers and their relation to the Divine Proportion, we will look at the number system we use today and then at the teachings of Pythagoras.

Our Number System

Our number system comprises different types of numbers. The most basic of these are natural numbers, integers, rational and irrational numbers, and real numbers.

Natural numbers are the numbers 1, 2, 3.... We use them for counting and for labeling the terms of a sequence, and they obey simple laws that are familiar to most of us. All natural numbers are integers.

Integers, or whole numbers, include zero and the negative numbers: ... -3, -2, -1, 0, 1, 2, 3.... The number 0 is only a placeholder with no numerical value of its own, and helps us to determine the difference between 10, 100, and 1000. Integers also obey simple laws. All integers are rational numbers.

Rational numbers are numbers that can be expressed as a ratio of two integers, but the denominator cannot be 0: -2/1, 1/2, -5/7, -121/457. They can be expressed either as fractions (7/9) or as decimals (0.53) with predictable patterns. All rational numbers are real numbers, meaning they are not imaginary.

Irrational numbers cannot be expressed as a ratio of integers and their decimal expansion does not terminate: $\sqrt{2}$, $\sqrt{3}$, $\sqrt{5}$. With rational numbers we see fractions like 1/11 which equals 0.090909..., a repeating decimal that is rational by definition. With irrational numbers we see numbers such as $\sqrt{2}$ which equals 1.41421356237..., a decimal that never repeats. Φ is also an irrational number. All irrational numbers are real numbers.

Real numbers (rational and irrational) are defined as numbers that are points on an infinite number line.

Thirteenth-century miniature from *The Life of the Infant Jesus*

The use of plus and minus symbols (+ and -) first appeared in text in the 14th century when Nicole d'Oresme (1323–1382) used a figure which looks like a plus symbol as an abbreviation for the Latin *et* (meaning "and") in *Algorismus proportionum*. However, the symbols were in use long before they appeared in mathematical texts. We know they were painted on barrels to indicate whether or not the barrels were full. Robert Recorde (ca. 1510–1558), who invented the modern sign of equality, did not use them in his arithmetic.

The equal symbol (=) was not used until 1557 when a Welshman named Robert Recorde, who first introduced algebra to England, decided to use a symbol, composed of two parallel lines, to avoid having to write over and over again "is equal to." The symbol didn't catch on right away, as some people preferred a different equal symbol (| |), while others used the abbreviation ae or oe (for the Latin *aequalis* or "equal") into the 1700s.

Number Games

THE PROPERTIES AND WORKINGS OF NUMBERS OFTEN SEEM QUITE MAGICAL.

285	For example, choose any three digit number whose ones and hundreds digits are different.
582	Reverse the order of the digits.
582 - 285	Subtract the smaller from the larger number.
=297	The result will always have a 9 as the tens digit and the other two numbers will always total 9.
792	Now reverse the digits of the result.
792 + 297	Add these two numbers.
= 1089	The result will always be 1089.

$$1+2+1=2^2$$
$$1+2+3+2+1=3^2$$
$$1+2+3+4+3+2+1=4^2$$
$$1+2+3+4+5+4+3+2+1=5^2$$

$$1^2=1$$
$$11^2=121$$
$$111^2=12321$$
$$1111^2=1234321$$
$$11111^2=123454321$$
$$111111^2=12345654321$$
$$1111111^2=1234567654321$$
$$11111111^2=123456787654321$$

1
1 1
1 2 1
1 3 3 1
1 4 6 4 1
1 5 10 10 5 1
1 6 15 20 15 6 1
1 7 21 35 35 21 7 1
1 8 28 56 70 56 28 8 1

THE PASCAL TRIANGLE is named after Blaise Pascal (1623–1662), although it had been described centuries earlier in 1303 by Chinese mathematician Yang Hui and the Persian astronomer-poet Omar Khayyám (1048–1123). In China it is known as the Yanghui triangle. Each term of the Pascal triangle is the sum of the two numbers above that flank it.

Quite recognizable as the forerunner of today's alphabet are the letters carved on the belly of this rooster-shaped vessel from the 7th century B.C.E. that was possibly an ink stand. When the Greeks came in contact with the Phoenicians (in about 800 B.C.E.) they borrowed their symbols to make their own alphabet. The Phoenician alphabet used more consonants than the Greeks needed for their language, so they used the extra signs for vowel sounds. This was an improvement over the Egyptian and Phoenician systems, because they could combine both consonants and vowels to form any sound they wanted. This alphabet was later adopted by the Romans, who developed it into much the same form as we use today.

Our Written Number System

About 3,700 years ago, some Semitic-speaking people of the Sinai (who later became the Phoenicians) were taken as workers or slaves under the Egyptians. The Egyptians had a complicated system of writing that was comprised of several hundred hieroglyphic symbols. The ancient Phoenicians adopted twenty-two of these symbols to write down sounds in their own language. Their system was later borrowed by the Greeks.

By the time Greek civilization began to flower in 500 B.C.E., the Greeks had a simple number system based on the decimal and comprised of the twenty-seven letters of their alphabet. The first nine letters represented the digits 1 through 9, the second nine letters represented the tens, and the last nine letters represented the hundreds. They used special signs to note numbers higher than 900. The Greeks did not have a zero and their numbers were nonpositional, meaning that they did not use columns as we do to give numbers their value. The great number of different symbols used for numbers made the system quite cumbersome to use, and calculation required a great deal of skill. We can barely imagine what it must have been like.

Pythagoras, First Pure Mathematician

Pythagoras (ca. 580–500 B.C.E.) was the first of the great teacher/philosophers of ancient Greece; perhaps because the Pythagorean Theorem has been attributed to him, he is also one of the most well known. Pythagoras was born shortly after

The School of Athens by Raphael (1483–1520)

Raphael's fresco attempts to represent the diverse pursuits of human reason by portraying some of the most well-known Classical thinkers in one mythical place and at one mythical time. At the center Plato and Aristotle walk through the school engaged in a dialogue.

The men in the picture include Pythagoras and Euclid; Socrates with his students Xenophon, Alcibiades, and Diogenes; Parmenides and his disciple Zeno; and thinkers from a much later time such as Epicurus. Also in the picture are Zoroaster from ancient Persia, who lived long before the time of Athens, and Averroes, the Islamic commentator on Aristotle who lived many centuries later.

Solon had come to power. A philosopher, mathematician, and founder of the Pythagorean brotherhood, he is often described as the first pure mathematician. His teachings greatly influenced the work of Socrates, Plato, and Aristotle, the three giants of Greek thought, who between them laid the philosophical foundations of Western culture.

Pythagoras belongs to the mystical wisdom tradition as much as he does to the mathematical one. He was a near contemporary of Buddha, Confucius, Mahavira, Lao-tzu, and probably Zoroaster; and although he left no writings behind, we do have some information about him both from legend and from people who wrote about him. We can certainly deduce that he was a much-revered figure among his followers, for the events of his life tend to be idealized and some of his later biographers portray him as a god-like figure. We also know that he founded a religious community and that the members of the community were required to keep certain teachings from the uninitiated.

Teachings of Pythagoras

The character of what is known as Pythagoreanism—the teachings that were developed by his students—is based upon ideas that include the metaphysics of number and a conception that reality (including music and astronomy) is, at its deepest level, mathematical in nature. The Pythagoreans devoted themselves to astronomical and geometrical speculation that combined a rational theory of numbers with a mystical numerology.

Their speculations on number and proportion led to an intuitive feeling about the *harmonia* ("fitting together") of the *kosmos* ("the beautiful order of things"). Their application of the

PYTHAGORUS: TEACHER, PHILOSOPHER, MATHEMATICIAN, MYSTIC

A detail of the painting by Raphael shows Pythagoras writing (although he did not) while a youth at his feet displays a panel on which there is a drawing of the *Tetraktys*. Pythagoras saw in the geometry of musical harmony a key to the order of the cosmos that he called the "harmony of the spheres."

PYTHAGORAS'S FATHER, MNESARCHUS, was a merchant who came from Tyre, a town on the Mediterranean coast of southern Lebanon. His mother came from Samos, a small Greek island in the Aegean Sea. There is a story that Mnesarchus brought corn to Samos during a time of famine and that the family was granted citizenship there as a mark of gratitude. As a child Pythagoras traveled widely with his father and was well educated.

Legend has it that when he was between eighteen and twenty years old Pythagoras met a famous sage of that time named Thales of Miletus Thales advised him to travel to Egypt to study their great secrets and to learn more about mathematics and astronomy from the priests there. In about 535 B.C.E., Pythagoras went to Egypt and the accounts of his time there suggest that he visited many of the temples and took part in discussions with the priests. According to one historian, Pythagoras was refused admission to all the temples except the one at Diospolis where he was accepted into the priesthood after completing strict admission rites.

In 525 B.C.E. Cambyses II, the king of Persia, invaded Egypt. Polycrates (the tyrannical ruler of Samos) abandoned an alliance that he had with Egypt and sent forty ships to join the Persian fleet against the Egyptians. After Cambyses won the battle Pythagoras was taken prisoner and taken to Babylon. Once freed he traveled east and studied with Zoroaster in Persia and possibly traveled to India to study with teachers there.

In about 520 B.C.E. he returned to Samos. Shortly after his return he made a journey to Crete in order to study the system of laws there. He then went to Croton and founded a religious community called "the semicircle" with an inner circle of followers known as *mathematikoi*. The *mathematikoi* lived permanently in the community where they were taught by Pythagoras. They had no personal possessions and were vegetarians. At length, Pythagoras and all his followers were banished from Samos by an opposing party, and he and his followers moved to Metapontum, a city in southeastern Italy that had been settled by Greeks two centuries earlier. It was there that Pythagoras died.

STORIES ABOUT PYTHAGORAS

ONE YEAR, WHEN PYTHAGORAS was on his way to Olympia for the athletic games, he met with a group of friends and fell into a discussion about prophecies, omens, and divine signs. He took the position that those who are attuned to their calling will receive messages from the gods. At that moment an eagle flew over his head and then turned, descended, and perched on Pythagoras' arm.

PASSING OVER THE RIVER CASUS near Metapontum with a group of his followers, Pythagoras paused on the bridge to pay his respect to the spirit of the river. In a distinct and clear voice, in the hearing of all, the river responded, "Greetings, Pythagoras!"

BIOGRAPHERS OF PYTHAGORAS have reported that during a single day he was present in Metapontum in Italy and in Tauromenium in Sicily, instructing disciples in both places. These cities are separated by some 200 miles of land and sea. Some say he was able to travel this distance by means of a sacred, golden dart given to him by Abaris the Hyperborean.

THE INVENTION OF THE WORD "philosophy" is attributed to Pythagoras. He was once asked, "Are you wise?" and he is said to have answered, "No, but I am a lover of wisdom." The Greek work *philo* means love and the word *sophia* means wisdom.

Another time he was asked, "What is philosophy?" and he is said to have given the following answer:

> *Life is like a gathering at the Olympic festival, to which, having set forth from different lives and backgrounds, people flock for three motives: To compete for the glory of the crown, to buy and sell, or as spectators. So in life, some enter the services for fame and others for money, but the best choice is that of these few who spend their time in the contemplation of nature and as lovers of wisdom.*

Pythagoras, the First Philosopher, an idealized contemporary illustration

ONCE, DURING A TRIP from Sybaris to Croton, Pythagoras happened to meet a group of fishermen as they drew up their nets filled with fish. As an amusement, he told them he knew the exact number of fish they had caught. The fishermen declared that if he was correct, they would do anything he said.

After the fish had been counted and it was found that he had predicted accurately, Pythagoras' request was simply that they return the fish alive to the sea, which they proceeded to do, and although they had remained out of the water for some time, not one fish had died. Pythagoras paid the fishermen the price of their fish, and continued on toward Croton.

tetraktys (from the Greek, *tetras*, four) to the theory of music revealed a hidden order in sound. Pythagoras referred to the "music of the heavens," which he claimed to hear, and came up with the idea that the distances of the heavenly bodies from the Earth corresponds to musical intervals—a theory that, under the influence of Platonic conceptions, resulted in the famous idea of the "harmony of the spheres." The Pythagoreans also investigated five solid figures, called the mathematical solids, a concept that was further developed by Plato and Euclid (and eventually Kepler) into "cosmic figures."

Some of the essential beliefs that Pythagoreans held were:

At its deepest level, reality is mathematical in nature.
Philosophy can be used as a tool for spiritual understanding.
The human soul can be experienced as in union with the divine.
Certain symbols have a real mystical significance.

Legend has it that in about 500 B.C.E. the Pythagorean philosopher Hippasus totally upset his contemporaries' worldview with the geometric proof that the ratio between the side and the diagonal of the rectangle cannot be expressed in integers. In other words, he proved the existence of irrational numbers. Apparently this happened while he was at sea with some comrades. The story says that upon notifying his comrades of his great discovery he was immediately thrown overboard. The logical truth of what he proved, it seems, was just too much to bear.

But it is hard to keep a secret, and the reality that Hippasus pointed to was not easy to dismiss. When the ancient Greeks were confronted with the fact of irrational numbers they called them "unnameable" or "speechless," because they were numbers

Pythagoras pointing to a globe on a Roman coin

Diogenes Laertius (3rd century C.E.), biographer of ancient philosophers, said about Pythagoras that he claimed to have been Hermes' son… "and Hermes told him he might choose any gift he liked except immortality; so he asked to retain through life and through death a memory of his experiences. Hence in life he could recall everything, and when he died he still kept the same memories."

Another story is told by Greek historian Diodorus (1st century B.C.E.) who writes:

> *They say that once when he was staying at Argos, he saw a shield from the spoils of Troy nailed up, and burst into tears. When the Argives asked him the reason for his emotion, he said that he himself had borne that shield at Troy when he was Euphorbus. They did not believe him and judged him to be mad, but he said he would provide a true sign that it was indeed the case: on the inside of the shield there had been inscribed in archaic lettering 'Euphorbus.' Because of the extraordinary nature of his claim, they all urged that the shield be taken down—and it turned out that on the inside the inscription was found.*

Pythagorus' Theorem: Mathematical Elegance

The Pythagorean Theorem

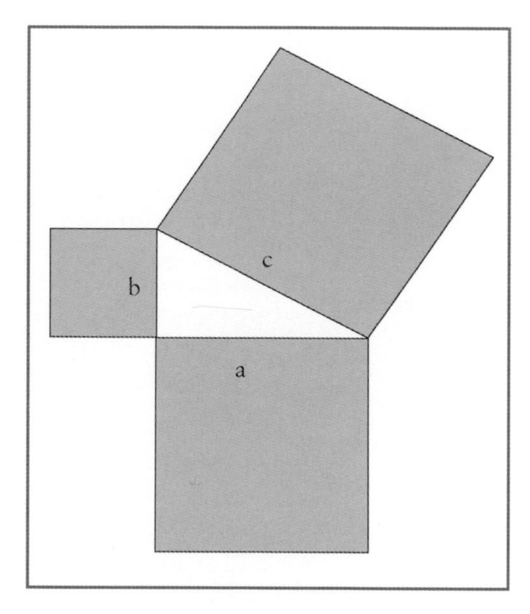

Pythagoras is perhaps best known as the author of the Pythagorean Theorem, a theorem that had been worked out long before in Egypt, Babylon, Japan, and India. The theorem states that for a right triangle with legs a and b and hypotenuse c, $a^2 + b^2 = c^2$.

Many different proofs exist for this theorem. The earliest known proof is a Chinese one that is illustrated in *Zhoubi suanjing* (The Arithmetical Classic of the Gnomon and the Circular Paths of Heaven).

Pythagorean Triples

It is possible to form right triangles with sides of whole numbers. The most famous of these is the 3, 4, 5 triangle. There are an infinite number of such Pythagorean triples, as they are called (5, 12, 13 and 7, 24, 25, for example).

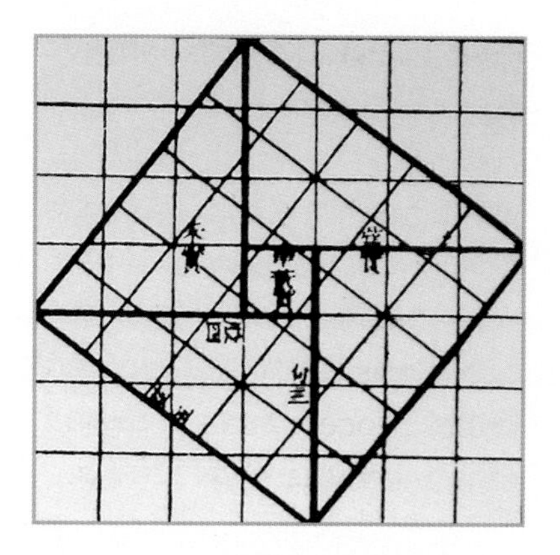

Proof of the Pythagorean Theorem is illustrated in the Chinese *Zhoubi suanjing* (above) and relies on the 3, 4, 5 triangle. Its proof can also be realized by studying the following diagrams:

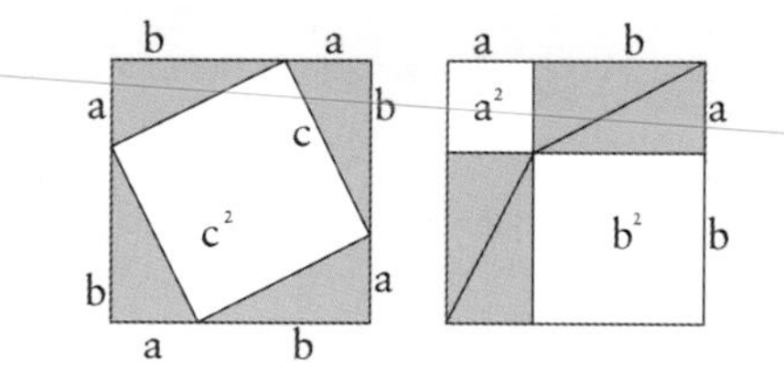

The area of the unshaded regions in each square are equal.

OTHER PROOFS OF THE PYTHAGOREAN THEOREM

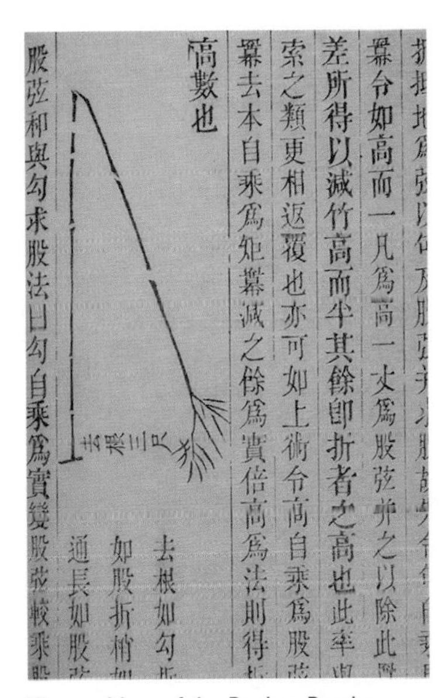
冪令如高而一凡爲高一丈爲股弦并之以除此
差所得以減竹高而半其餘卽折者之高也此率
索之類更相返覆也亦可如上術令高自乘爲股
冪去本自乘爲矩冪減之餘爲實倍高爲法則得
高數也

股弦和與勾求股法曰勾自乘爲實鏝股弦較乘

The problem of the Broken Bamboo comes from Yang Hui's *Xiangjie jiuzhang suanfa* (1261). The bamboo, which forms a natural right-angled triangle, is discussed as an example of the properties of such triangles in a wide range of problems.

These Babylonian clay tablets from the 2nd millennium B.C.E. explain how to calculate the areas and dimensions of various triangles, showing a familiarity with their properties.

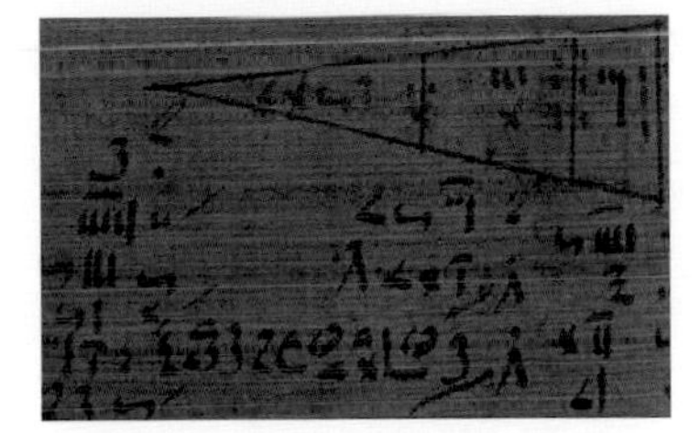

Problem number 56 in the Rhind Papyrus gives an equation to find the angle of the slope of a pyramid's face.

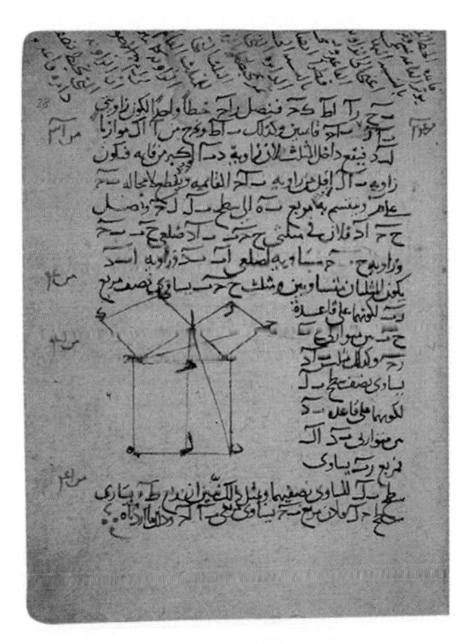

In a 13th century Arabic work titled *Euclid's Geometry Expounded*, the Pythagorean Theorem is proved using Euclid's "windmill" diagram.

Hermes Trismegistus

This figure of Hermes is from a pavement mosaic in Siena Cathedral (ca. 1488). The inscription praises him as a prophet of Christ. Hermes Trismegistus was originally an Egyptian priest and was known to the Egyptians as The Thrice Greatest, Ruler of the Three Worlds, Scribe of the Gods, and Keeper of the Books of Life.

The Greek God Hermes, who the Greeks believed was a messenger between the gods and mortals, evolved out of a mixture of the Egyptian God Thoth and Hermes Trismegistus. Pythagoras claimed to be born of him.

Hermes, a son of Zeus, was a prankster and inventive genius from birth, and loved by all the Gods. At five minutes old he stole his brother Apollo's sacred cattle and put shoes on their feet backwards to make reversed prints. He sacrificed two of them and hid the rest in a cave. Then stretching cowhide around a turtle shell and the guts across it, he put two horns through the shell's leg holes and created the lyre.

that cannot actually be named as real numbers. They can only be described by the operation that determines them, such as "the square root of 2."

The discovery of irrational numbers laid the grounds for the enormous philosophical problem for Greek mathematicians and philosophers concerning the makeup of the universe. Finding it difficult to accept the fact that some lengths were incommensurable with rational numbers (could not be measured exactly on a line), they decided that numbers could not be associated with lengths. Unfortunately, this decision led to a division between arithmetic and geometry that was not reconciled until the time of René Descartes (1596–1650), whose work *La géométrie* describes the application of algebra to geometry. Descartes' work led to Cartesian geometry, which makes it possible to quickly and elegantly prove some difficult questions through the combination of algebra and geometry that could not previously be proved at all.

Using the Pythagorean Theorem, we can easily see that irrational numbers are necessary to its equation. If the sides of a right triangle are one (1) unit in length, then the theorem says that $1^2 + 1^2 = 2^2$. The length of the hypotenuse is $\sqrt{2}$. Therefore there really is a length that logically deserves the name $\sqrt{2}$.

The ancient Greeks did not have anything like the algebra we use today. Nor did they use numerals the way we do. All of their thinking was based on logical thought involving words and abstract diagrams. So the discovery of these incommensurables did more than disturb the Pythagorean notion of the world, it led to a great impasse in mathematical reasoning. This situation persisted until the time of Plato, when geometers introduced a

René Descartes

DESCARTES: PHILOSOPHER, MATHEMATICIAN, DOUBTER

RENÉ DESCARTES (1596–1650) was a French mathematician, scientist, and philosopher. He was one of the first to abandon long-held views of Aristotle regarding logic and saw the early development of a new science grounded in observation and experiment. Applying a system of doubt to the established systems of knowledge led Descartes to intuit his famous phrase "I think, therefore I am" (best known in its Latin formulation, "Cogito, ergo sum," though originally written in French, "Je pense, donc je suis.")

Descartes was sent at the age of eight to a Jesuit school, and because of his delicate health he was permitted to lie in bed until late in the mornings. This was a custom he always followed, and when he visited Pascal in 1647 Decartes told him that the only way he had found to do good work in mathematics and to preserve his health was never to allow anyone to make him get up in the morning before he felt inclined to do so.

One day when he was about twenty, Descartes was walking through the streets and saw a placard in Dutch that excited his curiosity. He stopped passersby, trying to find someone who could translate it for him into either French or Latin. A stranger, who happened to be Isaac Beeckman, the head of a local college, offered to do so if Descartes would answer it; the placard being, in fact, a challenge to the entire world to solve a certain geometrical problem. Descartes worked the problem out within a few hours, developing a warm friendship with Beeckman as a result.

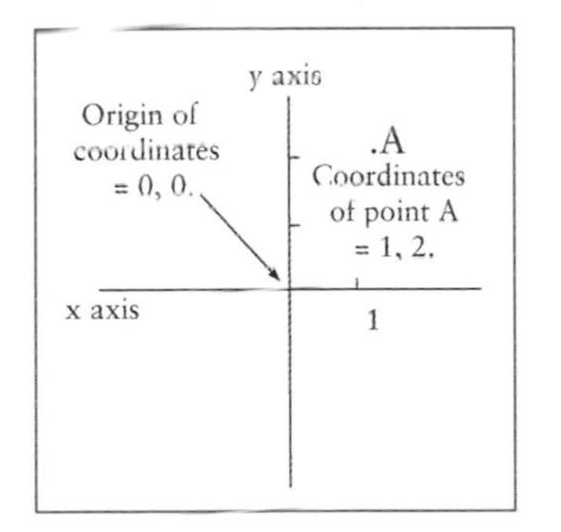

Cartesian geometry

CARTESIAN GEOMETRY

Cartesian geometry (also known as analytic geometry) connects arithmetic and geometry by establishing a set of coordinate axes against which geometric figures may be measured. In one-dimensional space, the only possible geometric figures are points and lines. Having chosen a point to be the origin, and a unit of length, we can assign a number to every point on a line by measuring it against a single x axis. In two-dimensional space, as illustrated, two numbers are needed to address each point. A distance along the x axis and a distance along the y axis meet in two dimensional space.

Descartes' linking of the ancient subjects of arithmetic and geometry enabled us to visualize arithmetic and algebraic ideas in geometric images and laid the foundations for a very fertile period in mathematics. It is second nature for us now to express a formula as a curve or some other shape in Cartesian space.

The ratio that Euclid describes says that the ratio of line AB to the larger part AC is the same as the ratio of the larger part AC to the smaller part CB. This gives a ratio of approximately 1.61803 to 1, which can also be expressed as $\frac{(1+\sqrt{5})}{2}$ or Φ.

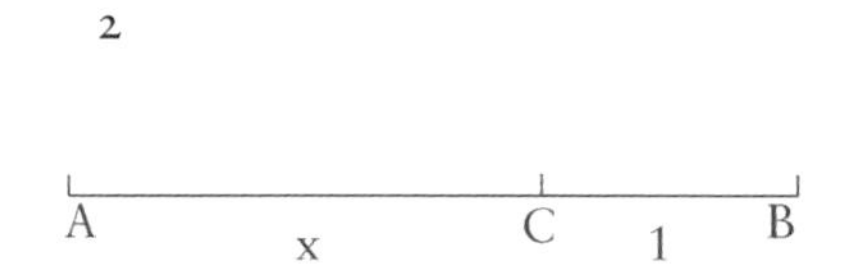

Let the length of AC be x and the length of CB be 1. Because the line has been cut into extreme and mean ratio, we know that the ratio of x to 1 is the same as x + 1 to x.

$$\frac{x}{1} = \frac{1+x}{x}$$

When we clear the denominators, we get $x^2 = x + 1$ or $x^2 - x - 1 = 0$. This is a quadratic equation, and there is a simple proof for x that gives two solutions for x:

$$\frac{(1+\sqrt{5})}{2} \quad \text{and} \quad \frac{(1-\sqrt{5})}{2}$$

The second value for x leads to a negative number which does not have meaning when associated with the length of a line segment. Therefore we will use only the positive solution and

$$\frac{(1+\sqrt{5})}{2}$$

is 1.6180339887.... This is the value of the Divine Proportion (Φ).

definition of proportion (ratio) that could account for the incommensurables. The main mathematicians involved were the Athenian Theaetetus (ca. 417–369 B.C.E.) and Eudoxus of Cnidus (ca. 390–340 B.C.E.), whose treatment of incommensurables survives as Book V of Euclid's *Elements*. It is in Book VI that we find Euclid's definition of the specific proportion that became known as the Golden Mean:

The line AB is divided in extreme and mean ratio by C if AB:AC = AC:CB.

The relationship that is established when a line is divided in this way is that the whole line is to the larger segment as the larger segment is to the smaller segment.

Pythagorean Doctrine of Emanations or the Science of Numbers

The Pythagorean doctrine of numbers touched upon harmony, geometry, number theory, and astronomy. Aristotle (384–322 B.C.E.) credited Pythagoras with the view that the principles of mathematics are the principles of all things. He writes in his *Metaphysics*:

> *...the so-called Pythagoreans, who were the first to take up mathematics, not only advanced this study, but also having been brought up in it they thought its principles* (archas) *were the principles of all things. Since of these principles numbers are by nature the first, and in numbers they seemed to see many resemblances to the things that*

exist and come into being—more than in fire and earth and water (such and such a modification of numbers being justice, another being soul and reason, another being opportunity—and similarly almost all other things being numerically expressible); since, again, they saw that the modifications and the ratios of the musical scales were expressible in numbers;—since, then, all other things seemed in their whole nature to be modeled on numbers, and numbers seemed to be the first things in the whole of nature, they supposed the elements of numbers to be the elements of all things, and the whole heaven to be a musical scale and a number.

In the worldview of medieval geometers the compass was seen as an abstract symbol for the eye of God. Its legs represented the rays of light and grace shining from heaven to Earth.

The system of principles behind numbers that the Pythagoreans used is complex, and owing to the fragmentary condition of existing Pythagorean records, it is difficult to arrive at an exact definition of the terms they used. However, the principles are beautiful and form the basis of many concepts of Western thought. An elementary understanding of them will explain much of the timeless and beautiful symbolism that is associated with the Divine Proportion, particularly through the pentad—the basis of the pentagram.

The ideas of the Pythagoreans were based on thoughts expressed in both words and symbols. They were not based on numbers as we know them today, although they incorporated many ideas we attach to numbers. The first ten numbers were seen as seed-patterns for all the principles of the cosmos. Beginning with a point, which is the essence of a circle, and using the geometers' tools of compass, straightedge, and pencil, the mathematical philosophers created a series of symbolic forms that mirrored their concepts of the universe.

VESICA PISCIS

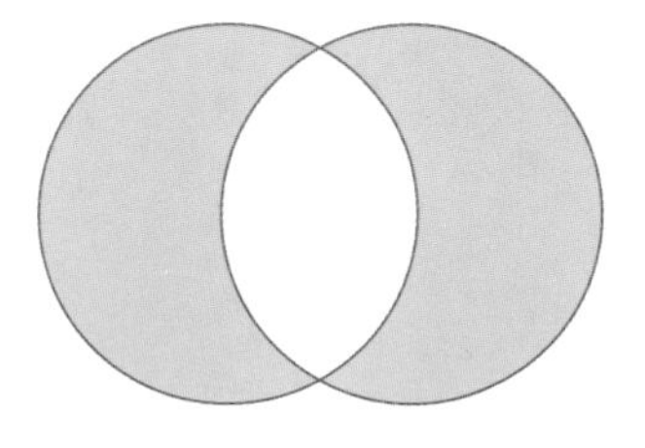

The form of the overlapping circles has appeared again and again, from time immemorial, in an ancient construction called the *vesica piscis*, Latin for "bladder of the fish." In the Christian tradition it is a reference to Christ and the "fish." It is called a *mandorla* ("almond") in India and was known in early civilizations of Mesopotamia, Africa, and Asia.

Nothing exists without a center, so they began with a point and around it they drew a circle. This represented the number one. This they called the monad. When the circle sees itself, it is mirrored, and then two circles exist (dyad), setting the stage for the generation of all the other numbers. Side by side the two circles overlap in such a way that they share their centers with each other. The shape that they share is called the *vesica piscis* (Latin for "fish's bladder"), and the first three shapes, as we shall see, to emerge from the *vesica piscis*, the triangle, the square, and the pentagon, form the relationships required to generate all further number-principles.

Diogenes Laertius (3rd century B.C.E) quotes earlier writers in his account of how the Pythagorean cosmology was constructed:

> *The principle of all things is the monad or unit; arising from this monad the undefined dyad or two serves as material substratum to the monad, which is cause; from the monad and the undefined dyad spring numbers; from numbers, points; from points, lines; from lines, plane figures; from plane figures, solid figures; from solid figures, sensible bodies, the elements of which are four, fire, water, earth and air; these elements interchange and turn into one another completely, and combine to produce a universe animate, intelligent, spherical, with the earth at its centre, the earth itself too being spherical and inhabited round about. There are also antipodes, and our 'down' is their 'up'.*

MONAD

The circle is the parent of all subsequent shapes. The Greek term for the principles represented by the circle was monad, from the root *menein*, "to be stable," and *monas*, or "Oneness." Ancient mathematical philosophers referred to the monad as The First, The Seed, The Essence, The Builder, and The Foundation. They also called it Unity.

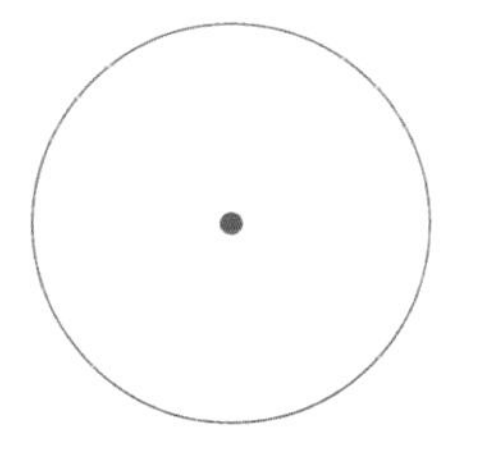

The relation of the monad to other numbers can be seen in the metaphor of simple arithmetic: Any number when multiplied by one remains itself (three times one equals three), the same is true when one divides into any number (five divided by one equals five). One, monad, preserves the identity of all it encounters.

The Pythagoreans believed that nothing exists without a center around which it revolves. The center is the source and it is beyond understanding, it is unknowable, but like a seed, the center will expand and will fulfill itself as a circle.

When mathematical philosophers noticed that no matter how many times one is multiplied by itself the result is always one, a significant question was raised: How does one become the many? Their answer was... as a reflection. The circle replicates itself by contemplating itself. This process is carried further in geometry as the birth of the line that connects their centers.

Unity, oneness, and the source are all symbols that emerge from monad.

The Tibetan Wheel of Life

Mystical Wheel by Fra Angelico (ca. 1400–1455)

Dyad symbolizes the polarities that interact to generate the world.

The Churning of the Sea of Milk, 19th century Hindu painting

The union of Upper and Lower Egypt is symbolized by the two figures of Hapy, God of the Nile flood.

Dyad

The principle of "twoness" or "otherness" was called dyad by the Greek philosophers. They referred to the dyad as "audacity," implying a boldness in separation from the original wholeness and "anguish" due to a yearning to return to oneness. Dyad or duality was also called "illusion." The principle of the dyad is polarity; it occurs everywhere and is at the root of our notion of separateness from each other, from nature, and from our own divinity.

The Greeks observed a paradox about the dyad: While it appears to separate from unity, its opposite poles remember their source and attract each other in an attempt to merge and return to the state of unity. The dyad simultaneously divides and unites, repels and attracts, separates and returns.

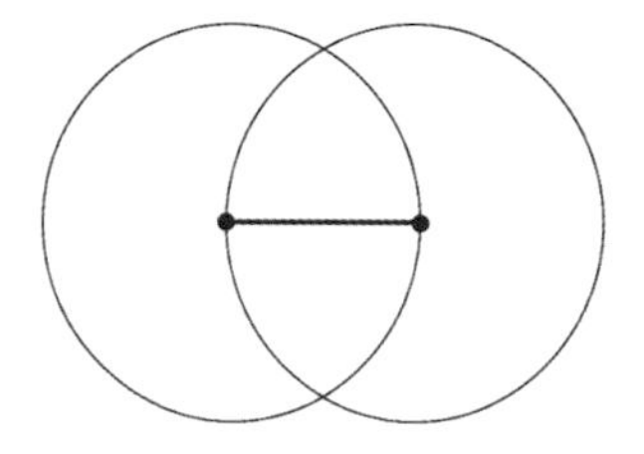

In the metaphor of arithmetic, the dyad reveals itself as the door between the One and the Many, between monad and all other numbers. The Pythagoreans saw in the ancient symbolic drawing of the *vesica piscis* a passageway to the journey of spiritual self-discovery. Its vulva shape had long been associated with fertility and the Divine Feminine and their interpretation of the symbol relates the spiritual journey to the passage of birth.

TRIAD

"One" and "two" were considered the parents of all the other numbers. The triad (symbol for three) is the firstborn, the eldest number. Its geometric expression, the equilateral triangle, is the initial shape to emerge from the *vesica piscis*, and it is the first of the Many.

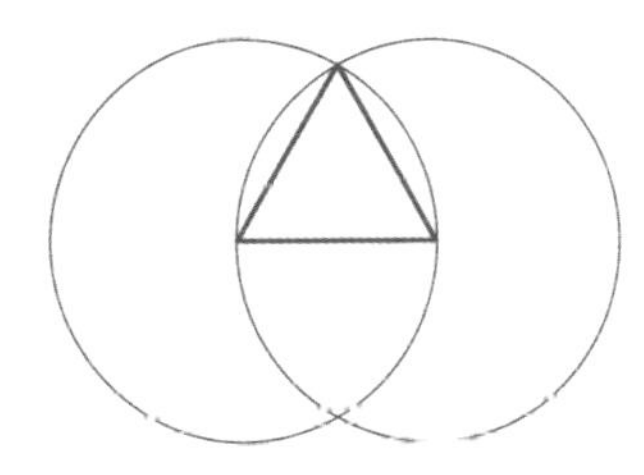

In comparison with the circle, which encloses the greatest area within the smallest perimeter, the triangle encloses the smallest area within the greatest perimeter. If we take a piece of string and tie its ends together we will find the truth of this.

> *The Triad has a special beauty and fairness beyond all numbers, primarily because it is the very first to make actual the potentialities of the Monad.*
>
> IAMBLICHUS (CA. 250–330), GREEK NEOPLATONIC PHILOSOPHER

Three is in a unique position. It is the only number to equal the sum of all terms below it (three equals two plus one) and the only number whose sum with those below equals their product (one plus two plus three equals one times two times three).

Ancient mathematical philosophers referred to the triad as prudence, wisdom, piety, friendship, peace, and harmony. To them, the shape of the triad was a statement about relationship and

The principles of trinity appear throughout myth and religion representing a relationship of mind, body, and spirit; birth, life, and death; past, present, and future.

The Holy Trinity by Andrei Rublev (ca. 1370–1430)

As Hindu Creator, Sustainer, and Destroyer, Brahma, Vishnu, and Shiva perform the Dance of Bliss.

The Egyptian God, Osiris, murdered by Seth, gives his divine sperm, the source of life, to his consort, Isis.

The stability of the triangle is used by builders in this detail from an illuminated chronicle depicting the construction of the great church of Bern by Diebold Schilling (15th century).

balance. As the centers of the two circles of the dyad repel and tug at each other, a reconciling third point occurs naturally above the place where the circles cross.

The archetype of the triad is of a relationship between opposites that unites them and brings them to a new level in their manifestation. No enduring resolution of any kind is possible without three aspects, two opposites and a neutral balancing, arbitrating, or transforming factor. Knowing how to choose the third element means the difference between a conflict's resolution and its perpetuation.

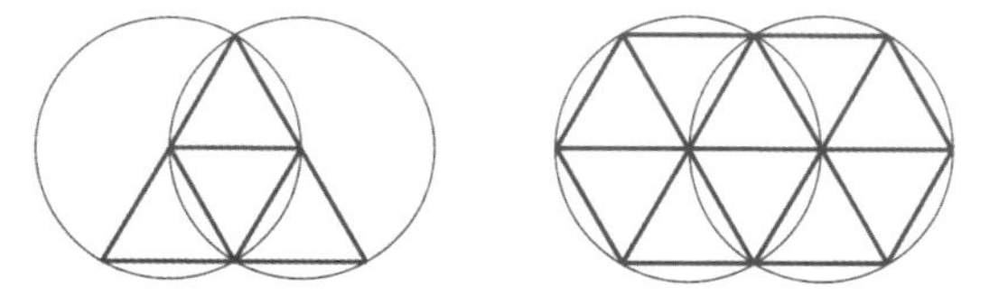

The triangle can be extended beyond the *vesica piscis* by extending the lines through the centers until they reach the opposite sides of the circles. When these points are connected with a horizontal line, a larger triangle appears. Further connections and further extensions result in a profound harmony.

The triangle is the only polygon structure that is rigid by virtue of its geometry. Its stability and strength are unmatched by any of its parts which by themselves do not have these properties. The three lines, unjoined, are meaningless. They are enhanced by the efficiency, balance, visual appeal, and symbolism of the triangle.

The word trinity derives from "tri-unity" or "three as one" and the triangle is the world's preeminent symbol of divinity. The ancient gods of the Hindu religion are called Trimurti (Sanskrit for "one whole having three forms").

TETRAD

In order to draw what next emerges from the *vesica piscis* we must use the faculty of logical thought to arrive at its construction. There are many ways to approach the problem of how the tetrad (symbol for four) emerges.

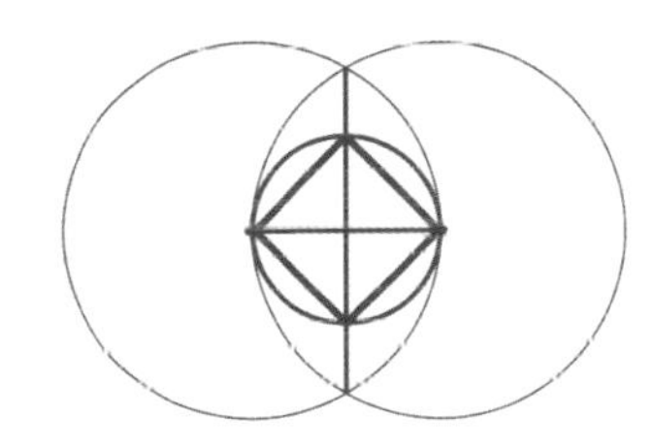

The simplest and most elegant is done by drawing a horizontal and a vertical line connecting the centers and the intersecting points of the two circles. When a circle is drawn along a line that connects the two centers there is a perfect square sitting within.

There are four ways to look at any three-dimensional structure—as points, lines, areas, and volume. The Greeks noticed that four is the first number formed by the addition and multiplication of equals (four equals two plus two and also two times two) and so four was considered the first even number and the first "female" number. To the Pythagoreans the even-sided square represented justice, because it is the first number divisible every way into equal parts (four equals two plus two or one plus one plus one plus one).

Four is associated with wholeness and completion, the four elements, the seasons, and the ages of man.

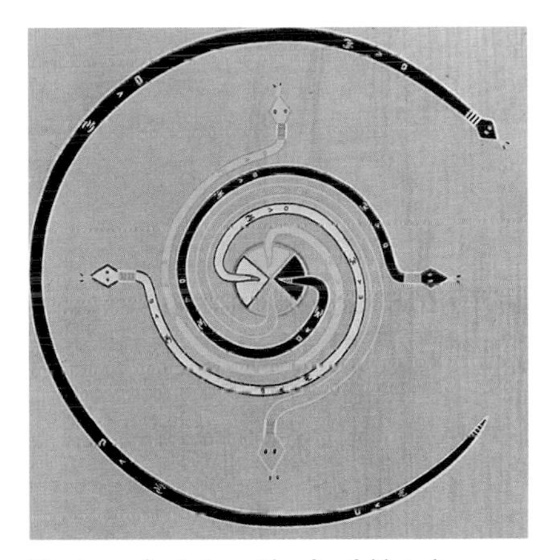

Navajo sand painting with a fourfold circle symbolizing integration

The mosque of Ibn Tulin in Cairo is aligned to the four directions.

The pentad is a symbol for growth.

PENTAD

From the development of the monad's point and the dyad's line there emerged the triad's surface and then the tetrad's three-dimensional volume. The pentad represents the next level of cosmic design by introducing the symbol of life itself.

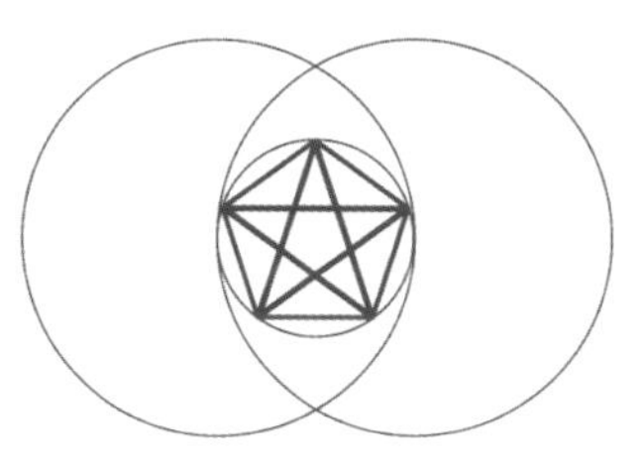

Seen most obviously as the five-pointed star, the pentad (symbol for five) shows up in many different places—from the five fingers on our hands and toes on our feet to its use in magic as a tool to ward off evil and as a symbol of power and invulnerability.

The pentad was so revered in early societies that its construction was kept secret. The Pythagoreans used it as a secret sign to recognize one another. They had studied its principles in geometry and nature and they also knew of its effect on the human psyche. They saw in its application a knowledge that could be misused, and although it was a teaching tool for a thousand years, it was guarded and only taught orally. It was not written about by the craft guilds who used its symbolism in the design of the Gothic cathedrals; and it was not until 1509, when Leonardo da Vinci's teacher Luca Pacioli published his book *Divina proportione*, that the method of its construction and unique geometric properties were publicly revealed to artists and philosophers.

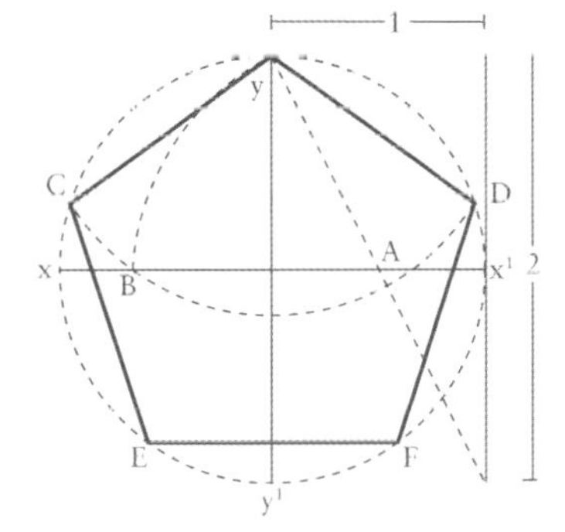

It is primarily from nature that we draw the pentad's symbolism and its relation to the Divine Proportion, for its geometry is found in abundance in vegetation. In seeds, leaves, and flowers there is very often a repeated fivefold rhythm.

But it is in the very act of regeneration, deeply related to Φ, that the most predominant principle of the pentad is symbolized. Here we see pentagram-shaped stars generating similar shaped stars in a repeating pattern of growth.

The Egyptian hieroglyph for "underworld" used the five-pointed star within a circle as a representation of the mystical nighttime place where the sun goes when it sinks below the horizon at dusk, the "underground womb." This carries over into a symbolic reference to the spiritual sleep from which humans must endeavor to awaken. The five-pointed star without a circle represented to the Egyptians a door or a teaching.

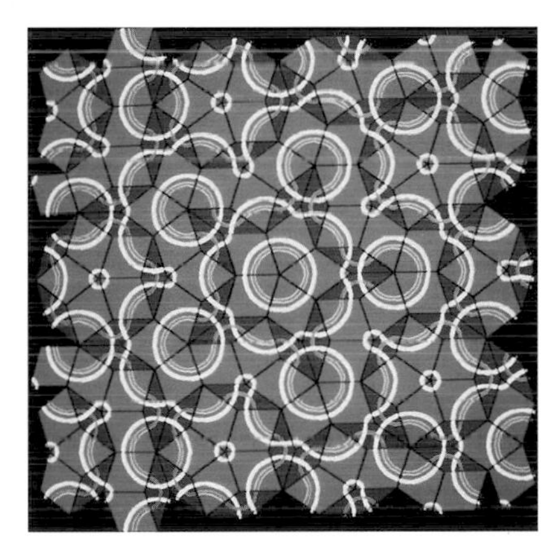

The present-day discoveries of Penrose tiling and quasicrystals that we will look at in Chapter 6 exhibit pentagonal symmetry.

Greek coin from 4th century B.C.E.

Hygeia from the Temple of Athena Alea in Tegea (350 B.C.E.)

In the way that leaves and flowers emerge it is easy to see the pentad's principle of self-similarity, or the smaller in the larger. The veins in a leaf reveal the branching pattern of the whole tree, each ridge of a fern's leaf models the whole leaf and the whole plant, each seed on a dandelion's head mimics the plant's head in miniature. In a crown of broccoli or cauliflower we see a reflection of the whole vegetable.

The pentad is perhaps best known to us in the symbol of the pentagram, the simplest form of a star shape that can be drawn with a continuous line. It is sometimes called the Endless Knot and is also called the Goblin's Cross, the Pentalpha, the Witch's Foot, the Devil's Star, and sometimes the Seal of Solomon (although this is more correctly attributed to the hexagram). The fivefold symmetries of the pentagram are inherent to the human body in our five senses: sight, hearing, smell, touch, and taste.

The earliest known uses of the pentagram have been found on ancient Babylonian pottery fragments that date back to 3500 B.C.E. In later periods of Mesopotamian art it was used in royal inscriptions and was symbolic of imperial power.

Pythagoreans considered the pentagram to be a symbol of perfection and labeled the five points or angles of the pentagram with the letters UGIEIA, combining EI into one letter. The letters labeling the angles are the first letters of Greek words for the elements:

U: Hudor = water
G: Gaia = earth
I: Idea = form/idea or Hieron = a divine, holy thing
EI: Heile = sun's warmth = heat
A: Aer = air

These letters were translated into the word *hygeia*, which literally translated means health, but carries the sense of soundness, wholeness, or a divine blessing. Hygeia is the Greek Goddess of health, and her name was a fairly common inscription on amulets. The apple, one of our symbols of health, when cut through its core, reveals a pentagram.

This famous ceramic painting from 540–530 B.C.E. shows Dionysus on a ship. It alludes to a myth of his abduction by Titans, who had originally ruled the Earth, and their transformation into dolphins.

It is possible that Pythagoras became acquainted with earlier symbolism of the five-pointed star during his travels to Egypt and Babylon and incorporated other meanings into the symbol. The early ancient Greek pentagram had two points up and represented the doctrine of *Pentemychos* that was written about by Pythagoras' teacher and friend Pherecydes of Syros. *Pentemychos* refers to the five recesses or chambers, also known as the *pentagonas*, and was the place where the first pre-cosmic offspring had to be put in order for the ordered cosmos to appear. Although locked away after the emergence and ordering of the cosmos, this place still continued to have an influence. It was described by Homer as "the subduer of both gods and men" and played a role in many early myths that saw the underworld as a source of wisdom.

15th century illustration of Christ and Satan

Early Christians used the pentagram to represent the five wounds of Christ and as a symbol for truth. It has also been used to refer to the star of Bethlehem. Emperor Constantine I, after gaining the help of the Christian church in his takeover of the Roman Empire, used the pentagram, together with a form of the cross, in his seal and amulet.

There have been times in which the pentagram was used as a symbol for the devil. This most likely began during the Christian Inquisition when the inquisitors believed that any

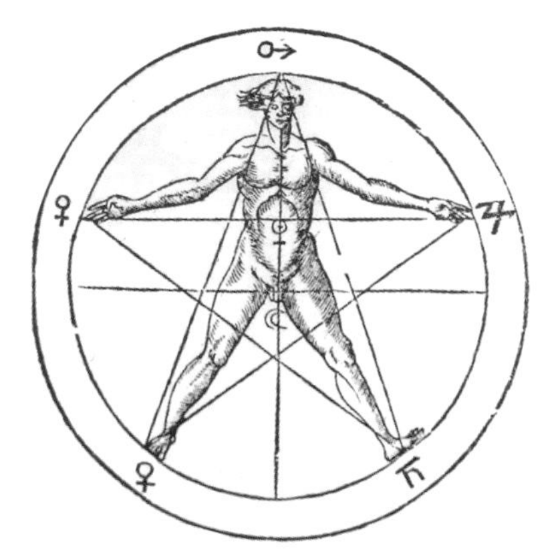

Image of a pentagram from Heinrich Cornelius Agrippa's *Libri tres de occulta philosophia* illustrating the symmetry of the human body. Agrippa (1486–1535) was a prolific Renaissance writer of esoterica.

In *The Sacrament of the Last Supper*, Salvador Dali (1904–1989) uses the pentagram symbolism.

non-Christian was obviously worshiping Satan, and the symbol became synonymous with Devil worship. In the purge of witches, the pentagram came to be seen as evil and was called the Witch's Foot.

When this interpretation is used the point usually faces downward, and the pentagram is thought to represent the face of a goat, confusing Satan with the Greco/Roman God Pan. The association of the pentagram with Satan may have been furthered by the symbolic association of the term Lucifer with Satan.

A circle around a pentagram contains and protects its core. The circle symbolizes eternity, infinity, and the cycles of life and nature. The center of a pentagram implies a sixth, formative element of love that exerts its own power from within.

In Goethe's *Faust* the pentagram hinders Mephistopheles from leaving a room.

> *Mephistopheles:*
> *I must confess, my stepping o'er*
> *Thy threshold a slight hindrance doth impede;*
> *The wizard-foot doth me retain.*
>
> *Faust:*
> *The pentagram thy peace doth mar?*
> *To me, thou son of hell, explain,*
> *How earnest thou in, if this thine exit bar?*
> *Could such a spirit aught ensnare?*

In the medieval romance of *Sir Gawain and the Green Knight*, a pentagram, or pentangle, was inscribed in gold on Sir Gawain's shield to symbolize the five knightly virtues of generosity, courtesy, chastity, chivalry, and piety.

A chivalrous knight from a 15th century tapestry

Then they showed him the shield of shining gules,
With the Pentangle in pure gold depicted thereon.
He brandished it by the baldric, and about his neck
He slung it in a seemly way, and it suited him well.
And I intend to tell you, though I tarry therefore,
Why the Pentangle is proper to this prince of knights.
It is a symbol which Solomon conceived once
To betoken holy truth, by its intrinsic right,
For it is a figure which has five points,
And each line overlaps and is locked with another;
And it is endless everywhere, and the English call it,
In all the land, I hear, the Endless Knot.
Therefore it goes with Sir Gawain and his gleaming armour,
For, ever faithful in five things, each in fivefold manner,
Gawain was reputed good and, like gold well refined,
He was devoid of all villainy, every virtue displaying
In the field.

Thus this Pentangle new
He carried on coat and shield,
As a man of troth most true
And knightly name annealed.

Decad heralds a new beginning and opens up the possibility of many.

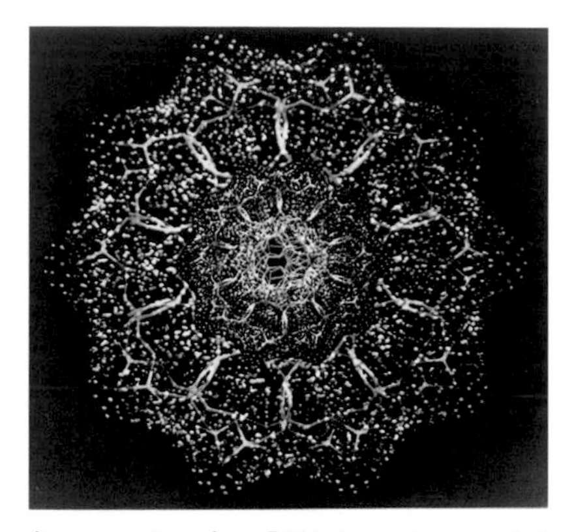

A cross section of our DNA shows the ten-sided form of decad.

DECAD

Based upon the appearance of the first four numbers all other numbers emerge. Step by step they approach the decad, or number ten. Step by step we move beyond the ordinary numerical interactions and geometric relationships and find a new beginning with the decad, the beginning of a journey into limitlessness.

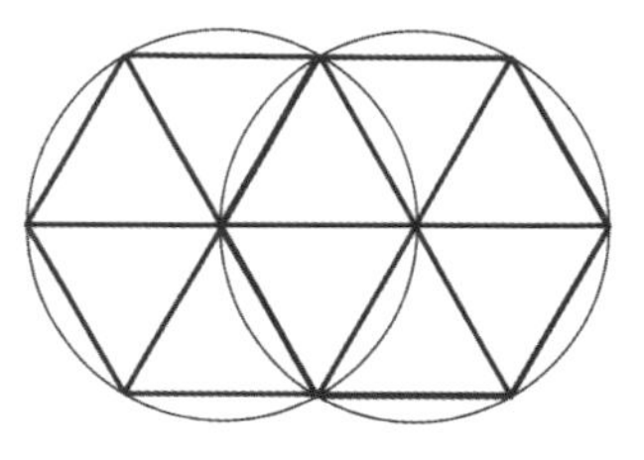

Like one and two, ten was not considered a number by the ancient mathematicians. Rather, it was seen as an assembly point and became a symbol of both world and heaven. Like our two hands with their ten fingers and the ability that gives us, the decad with its ten-ness comprises all that is necessary for understanding the construction of the universe. Since ten equals one times two times five it is the result of the interplay of monad, dyad, and pentad. Like the monad, any number multiplied by ten is similar to multiplying that number by one. The result is essentially unchanged, but the number is brought to a higher level and becomes an expanded version of itself.

> *Ten is the very nature of number. All Greeks and all barbarians alike count up to ten, and having reached ten revert again to the unity. And again, Pythagoras maintains, the power of the number 10 lies in the number 4, the tetrad. This is the reason: if one starts at the unit (1)*

and adds the successive number up to 4, one will make up the number 10 (1+2+3+4 = 10). And if one exceeds the tetrad, one will exceed 10 too.... So that the number by the unit resides in the number 10, but potentially in the number 4. And so the Pythagoreans used to invoke the Tetrad as their most binding oath: "By him that gave to our generation the Tetraktys, which contains the fount and root of eternal nature..."

(AETIUS I. 3.8)

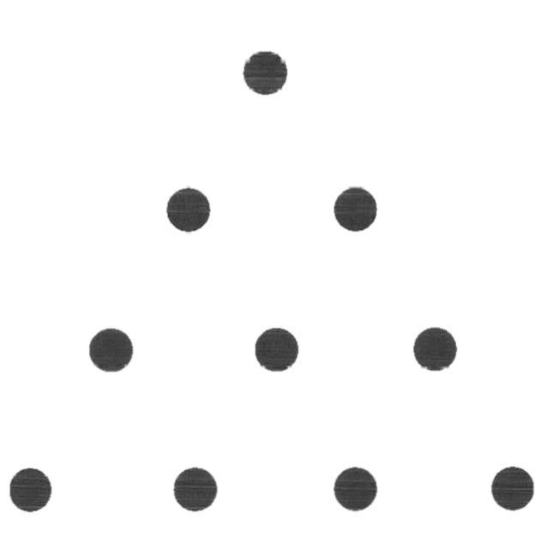

Ten pebbles (or points) are arranged in four rows to form an equilateral triangle. This creates the Tetraktys (from the Greek word for "fourfold"), a powerful diagram that was a metaphor for the way that numbers relate to the universe.

Four Elements: earth, water, air, fire

Three Principles: salt, quicksilver, sulphur

Two Seeds: sun, moon

One Fruit: world soul

Its four levels also represent increasing densities of the four elements, fire, air, water, and earth, as well as the numbers: one represents the point, two represents the line, three represents the surface, and four the three-dimensional form. The Pythagoreans also saw in this form the growth of a plant: from seed to stem, leaf, and fruit.

The Tetraktys also became a diagram for their discoveries in music. Pythagoras experimented with stretched strings of different lengths placed under the same tension and discovered the relation between the length of a vibrating string and the pitch of the note. The Tetraktys contains the symphonic ratios which underlie the mathematical harmony of the musical scale: 1:2, the octave; 2:3, the perfect fifth; and 3:4, the perfect fourth.

Thus the early philosophers found harmony in numbers—a harmony that was clearly reflected in the order of nature, art, science and sound. It was a mysterious harmony, not clearly understood, but beautiful and deeply symbolic.

Do not seek to follow the footsteps of the men of old: seek what they sought.

MATSUO BASHO (1644–1694), JAPANESE POET

Chapter 3

FIBONACCI AND THE FIBONACCI NUMBERS

The real voyage of discovery consists not in seeking new lands but seeing with new eyes.

MARCEL PROUST (1871–1922), FRENCH NOVELIST

THE DIVINE PROPORTION has been called divine because people have found it to be so. It is creative, regenerative, harmonious, and though it eternally approaches something unknown, it never arrives. Since the earliest of times the Divine Proportion has been used by artists, sculptors, musicians, and perhaps even poets as the harmonic basis for their works; and mathematicians, physicists, botanists, and all manner of scientific inquirers have been inspired by it to elicit nature's secrets. Each time it appears in some new and wondrous form, as it repeatedly does, we are amazed; and we hear the collective hum—does nature really have some secret code that can be understood by the human mind?

Frontispiece to a highly popular 16th-century text, the *Suanfa tongzong (General Source of Computational Methods),* shows the use of a counting board for mathematical calculations.

As we saw in the previous chapter, the Pythagoreans considered numbers to be the essence and principle of all things, the very elements out of which the universe was constructed. The system they developed helped them to understand the relationship of

OPPOSITE: "Takiyuddin in His Observatory at Galata" a 16th century illustration from *Shahanshanama (Book of the King of Kings)* by Lokman.

Athenian traders at work

A young man recites in front of his teacher, who follows the text on a papyrus roll.

human and divine nature and implied to them the existence of a somewhat secret code. The Pythagoreans were called *mathematekoi*, which means "those who study all." The word *mathema* is the root of the Old English *mathein*, "to be aware," and the Old German *munthen*, "to awaken." The truths the Pythagoreans sought were the universal truths of self-knowledge and their inquiries relied upon finding an application of these truths in the world around them. The Pythagorean philosophy was based on an attempt to describe the underlying harmony of existence and the nature of the perfect universe in numbers.

What they sought in numbers was something many humans have long sought. Observing principles from the natural world, creating man-made systems from abstractions, and then turning these principles into mathematical thought has led many an inquirer into dangerous new lands. Discoveries have many times been regarded as some type of heresy. Rigorous proofs have sometimes defied acceptable knowledge and pushed human consciousness to accept awkward realities. The Pythagoreans, truth-loving as they professed to be, were stunned to uncover the principles of irrational numbers that lay just under the surface of their perfect code.

Courageously, though, they rose to the challenge offered by their new discovery and laid a foundation for further mathematical principles to evolve. What finally unfolded in the years that followed their early inquiries is the wonderfully descriptive language of mathematics. As this language grew in its ability to express abstract principles, the Golden Ratio (which later became the Divine Proportion) slowly emerged from its roots. Originally used in ancient Egypt, the Divine Proportion was to be rediscovered in a surprising new form by a young man from Pisa.

Earliest Expressions of Divine Proportion

Is there indeed a secret code that reverberates throughout nature, science, and art, and to which humans intuitively respond? There is no doubt we would love to think so; and it may be true. The mystics of the world from the early days of the world's great river valleys certainly all point to some great harmony. Pythagoras claimed to have been able to hear it.

The Pyramids of Giza

The Great Pyramid at Giza is the largest of the group at Giza and is the burial place of King Khufu. Its remarkable beauty is based on a relationship of Divine Proportion.

When we ponder the building of the pyramids in Egypt, it is almost inconceivable to us, armed as we are with calculators, telescopes, microscopes, computers, and all kinds of devices to insure accuracy and alignment, that the people of that bygone era could not only conceive, but create, the monumental works that they did—well designed, and perfectly aligned structures of enormous scope and magnitude—without some perfect tool at their disposal. Although there is no way for us to know for sure how they achieved what they did, we all applaud the great genius that lies behind these works...and go on searching for some clue as to what tools they used.

Maybe they had no single tool, maybe their methods of calculation were extremely cumbersome and awkward, and maybe all they had was some inner sense of perfect balance, perfect harmony. Maybe they understood more perfectly than we, with modern knowledge at our fingertips, how to translate the inner laws about the relationship of things of which they were kinesthetically aware. Maybe theirs was an innately felt experience about the divine proportioning of the whole cosmos and all its interrelated parts. If they were unsure of their theory and wanted to prove it... they did!

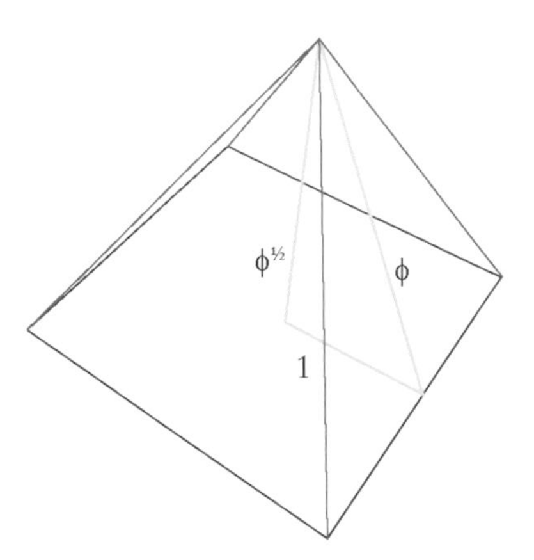

This diagram shows the mathematical relationship between the Great Pyramid's height, its face, and the Divine Proportion.

The Great Pyramid of Giza was the oldest of the Seven Wonders of the Ancient World and the only one that still survives. It is located in the burial grounds of ancient Memphis that today is part of Greater Cairo. The monument, which probably took about twenty years to complete, was built to serve as a tomb by the Egyptian pharaoh Khufu of the Fourth Dynasty around the year 2500 B.C.E.

When it was built, the Great Pyramid was 481 feet high. Over the years, it has lost thirty feet off its top, for it was covered with a casing of stones to smooth its surface and most of these have fallen away. The sloping angle of the pyramid's sides is 51 degrees and 51 minutes. Each side is carefully oriented with one of the cardinal points of the compass, north, south, east, and west. The horizontal cross section of the pyramid is square at any level, and each side measures 751 ft in length. The maximum error between side lengths is astonishingly less than 0.1%.

The Greek historian Herodotus reported that the pyramid was built in such a way that the area of each face would equal the area of a square whose side is equal to the pyramid's height. In other words, the relationship of the pyramid's faces to its height is a relationship of Divine Proportion.

No one can prove with any certainty that the architects of the Great Pyramid used this relationship; numerous books and papers have been written both proving and disproving the theory. It is certainly an interesting speculation, however, and given the visual beauty of the pyramid, we can be almost certain that ratios were considered in some form or other. There are many indications from early Greek writings that lead us to suspect the Egyptians had a profound understanding of universal laws and

THE GREAT PYRAMID OF KHUFU AT GIZA

The production of mudbricks is depicted in a tomb at Thebes.

Any statues of Khufu that existed in the temple complex have vanished. The only surviving image of the king who was buried at the Great Pyramid is this tiny statuette found at another site.

The structure consists of approximately two million blocks of stone, each weighing more than two tons. It has been suggested that there are enough blocks in the complex of the three pyramids to build a ten-foot-high, one-foot-thick wall around France. The area covered by the three pyramids can accommodate St. Peter's in Rome, the cathedrals of Florence and Milan, and Westminster and St. Paul's in London combined.

> *Twenty years were spent in erecting the pyramid itself: of this, which is square, each face is eight plethra, and the height is the same; it is composed of polished stones, and jointed with the greatest exactness; none of the stones are less than thirty feet.*
>
> HERODOTUS, *HISTORY*

Peasants Playing Bowls Outside a Village Inn by David Teniers the Younger (1610–1690)

Allegory of Good Government: Effects of Good Government in the City (detail) by Ambrogio Lorenzetti

The councilors of Siena commissioned this mural to decorate the town hall and praise the good effects of a well-ordered state. Up until quite recently, the most widely accepted models of the universe have suggested that it is both flat and infinite. Flatness in this case does not mean that the space is two-dimensional, but rather that it has no large-scale curves in it.

that they maintained an extraordinary body of knowledge. The development of Western philosophy that emerged in Greece during the 5th century B.C.E. was unquestionably inspired by teachings that came from the Nile Valley.

Though we have no real knowledge about the roots of the Egyptian experience, we know that the Greek development was based on the empirical beauty of logical thought. After the difficult problem of the incommensurables was presented to them by Pythagoreans, there was a period of perhaps 150 years of great expansion when the Persian and Peloponnesian Wars were fought and the Parthenon was built. This was the period known as the Classical period of Greek culture, and it was a period in which Greece enjoyed a cultural flowering and economic prosperity rarely seen in the world. The Golden Mean or Divine Proportion certainly appeared in works of art and architecture during this period.

The Dark Ages in Europe and the Appearance of Fibonacci

After the Roman Empire finally collapsed in 476, the Dark Ages hung over Europe for almost 1,000 years. Invading tribes coalesced into small regional kingdoms and the common people were bound to the land, dependent on landlords for protection and some semblance of justice. The momentary rise to power of Charlemagne in the 9th century brought a brief glimmer of order, but the onslaughts of the Vikings in the 10th century forced Europe back into the endless violence of war.

By the 12th century things had changed. Towns once again became the focus of civilization and their populations steadily grew along with education and a revival of interest in Greek science and philosophy. Invasions, migrations, and population decline had come to an end. The Roman Church was no longer God's kingdom in exile; it became the central institution of power. Italian merchant-bankers penetrated much of Europe and North Africa, and Italy slowly became a seat of power and authority, both economically and ecclesiastically.

Fibonacci was born into these times—Fibonacci who unknowingly made the next great contribution to a growing European understanding of the Divine Proportion with the introduction of a certain puzzling problem in one of his books.

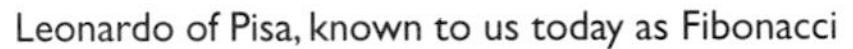

Leonardo of Pisa, known to us today as Fibonacci

When my father, who had been appointed by his country as public notary in the customs house at Bugia, established for the Pisan merchants going there, was in charge, he summoned me to him while I was still a child, and having an eye to usefulness and future convenience, desired me to stay there and receive instruction in the school of accounting. There, when I had been introduced to the art of the Indians' nine symbols, a remarkable teaching, knowledge of the art very soon pleased me above all else and I studied with them, whoever was learned in it, from nearby Egypt, Syria, Greece, Sicily and Provence, and practiced their various methods.

From Fibonacci's *Liber abaci*

The author of the famous Fibonacci series was an Italian mathematician, known to us today as Fibonacci (ca. 1170–1240). His real name was Leonardo or Leonardo of Pisa. His father (a business and government official) may have been named Bonacci, for Leonardo was posthumously given the nickname Fibonacci (for *filius Bonacci*, son of Bonacci), but the name could equally as well have come from his good nature (another meaning of the word Bonacci is "good nature.") In some manuscripts he refers to himself as Leonardi Bigolli Pisani where "Bigollo" means something like "good for nothing" or "traveler."

Born in Italy, Fibonacci was educated in North Africa where his father, Guglielmo, held a diplomatic post in what was then called Bugia (and today is called Bejaïa), a Mediterranean port in northeastern Algeria. His father's job was to represent the

Fibonacci talked with merchants and understood the needs of those who traded their goods in bazaars such as this.

merchants of the Republic of Pisa who were trading there. Fibonacci was taught mathematics in Bugia and was introduced to the nine figures of what was then known as Indian counting.

We do not know how widely the young Fibonacci traveled, so we do not know where or how he was introduced to the system; but he appears to have immediately grasped the enormous possibilities of its applicability.

Calculation of Numbers at the Time of Fibonacci

The number writing system in use at that time in Pisa, and universally throughout Western Europe, consisted of Roman numerals. Starting with "stick" numbers—I, II, III, IIII—they had added a half dozen letter symbols—V for 5, X for 10, L for 50, C for 100, D for 500, and M for 1,000. This system had the advantage of simplicity over the Greek system of Euclid, for only seven symbols had to be learned. Addition and subtraction were relatively simple with these numbers, but multiplication and division were a different matter.

Addition with Roman numbers can be done quite easily:

	M	CC	XX	III	1,223
Plus	M	C	X	II	1,112
equals	MM	CCC	XXX	V	2,335

But what would be a simple addition problem to us, grows complicated quite quickly in this system:

	CCLCVI
+	DCL
+	MLXXX
+	MDCCCVII

Translated into our own system this problem reads:

	266
+	650
+	1,080
+	1,807
	3,803

An astronomer with his astrolabe and two assistants, one with an abacus, from the Psalter of St. Louis and Blanche of Castile (13th century)

However difficult the Roman system might have been to those who used it, an ingenious device called the abacus existed alongside the written system and provided what the system did not—place value—the assignment of a value to every number, depending on its place in a row. Apart from the abacus, place value had no part in the Roman system. Oddly enough, the Babylonians had employed a place-value system, but for some unknown reason it had disappeared.

Using either the abacus or a written method, calculations that involved multiplication or division relied upon addition and

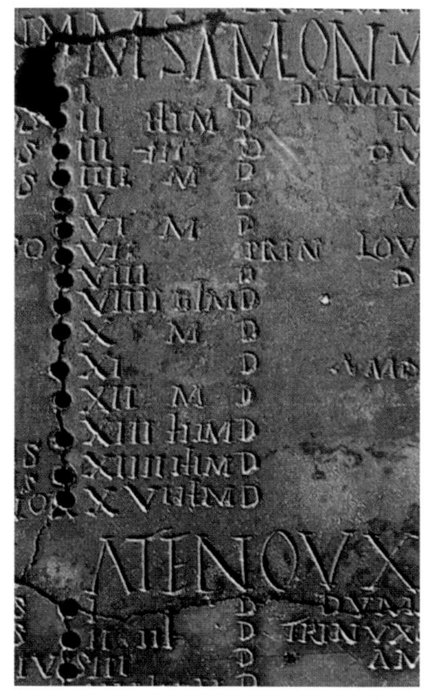

Roman number notation in an engraved calendar

subtraction for their execution. In order to multiply 24 x 3, twenty-four was added three times:

XXIIII
XXIIII
XXIIII

To divide 24 by 3, three was subtracted from twenty-four until nothing remained:

XXIIII -III = XXI
XXI - III = XVIII
XVIII - III = XV
XV - III = XII
XII - III = VIIII
VIII - III = VI
VI - III = III
III - III = nothing

Thus, after quite some time, it would become clear that there were eight threes in twenty-four.

One problem with the abacus was that the work vanished as it was performed, leaving no record of the process. There was no way to check a calculation except by doing it over again. In the time of Fibonacci's youth, the growing complexity of business transactions demanded a more sophisticated technique. A few Europeans (mostly in Spain) had become aware of the use of foreign numerals, but no one realized their importance. No one...except the young Leonardo, recently landed in Bugia, whose fascination with numbers and whose acute intelligence led him to grasp at all manner of knowledge.

THE ABACUS

THE WORD ABACUS MAY HAVE ORIGINATED from the Hebrew word for dust, *avaq*, for the ingenious instrument originated as a board sprinkled with sand on which tallies were made by hand and then erased. At some point, someone had the brilliant idea of drawing lines in the sand and dividing the board into columns—one for tens, one for hundreds, and one for thousands. The Romans refined the abacus by constructing boards marked with grooves in which pebbles or counters of some sort could be moved without falling into the wrong column. They also developed, or possibly borrowed from somewhere, a portable abacus—a frame with wires on which beads were strung.

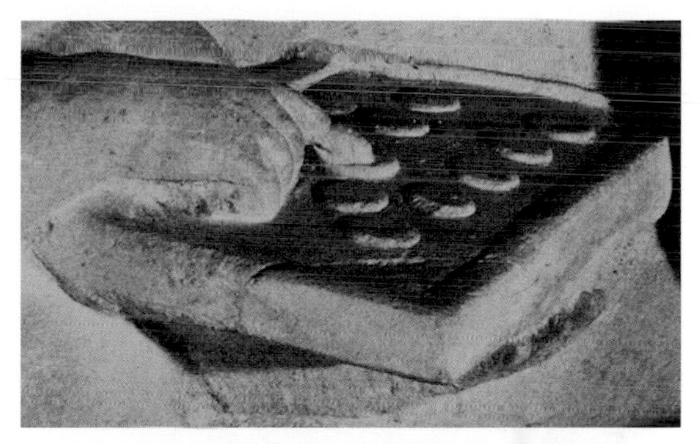

IN THIS DETAIL from a marble sculpture (1st century B.C.E.) the slave of a dying man uses an abacus to check that the bequests in his will do not outrun his estate.

A ROMAN CALCULATOR on this engraved seal holds a writing tablet in his left hand while he configures the abacus on the table with his right hand.

THIS DETAIL FROM AN AZTEC VASE found in Guatemala shows a seated man using an abacus to calculate the value of tax being collected in the form of cacao beans.

The fourteen-hundred-year history of Islam has seen magnificent contributions to science, medicine, poetry, art, and architecture.

As the Muslim realm expanded it brought an increasing diversity of cultures under its domain.

Advances in the Arab World

Twelfth-century Bugia was one of many flourishing centers of Islam. Five and a half centuries earlier a successful Arab businessman named Muhammad had preached a new religion in Mecca. Such was the appeal of this religion—based like Judaism and Christianity upon the idea of a single God—that within a few years Mecca, which was already a large terminal of the trans-Arabian trade route to India, became a center for pilgrim traffic. In their passionate and religious zeal to free themselves from their tribal rulers, Muhammad's followers developed into fearless warriors who rapidly overcame their enemies in Persia. They then extended their stronghold to lands that stretched across the entire Arabian Peninsula, up to the Caspian Sea, across North Africa, and into Spain.

Merchants and scholars ventured far and wide, bringing both silk and the secrets of paper making from China along the Silk Road. Baghdad and other Muslim centers grew in culture and in wealth, and because of the wide range of thought that travelers were exposed to, these centers became the meeting places of enormous intellectual curiosity. While the rest of Europe had fallen into the darkness of the Middle Ages and forgotten the writings of the Greeks, Muslim scholars began translating the great Greek masters into Arabic. In 830 Caliph al-Mamun founded the House of Wisdom in Baghdad as a center for knowledge. Special commissions were sent to Constantinople to copy Greek manuscripts so they could be translated into Arabic. Peace treaties with the Byzantines sometimes included provisions for the turning over of Greek manuscripts.

In the 9th century, the darkest of the Dark Ages in Europe, the Muslim world was reading Arabic translations of Aristotle's philosophical works, the astronomy of Ptolemy, Euclid's *Elements*, and the medical writings of Hippocrates and Galen. Several works that were lost in their original Greek were preserved (and enriched) for posterity in their Arabic translations.

Reaching into the Far East, traveling Arabs brought back from India the discovery that was to spark Fibonacci's imagination—a system of writing down numbers that would revive ancient mathematical skills—the systems of both the place-value holder and of zero.

Mathematical Developments in India

During the centuries after the great mathematical advances of the Greeks, when Europe's learning had spun into decline, there had been an inspired flowering in India, galvanized perhaps by the country's contact with the Arab world. The great astronomer-mathematicians Aryabhata (d. 550), Brahmagupta (d. 660), and many others took up algebra where the Greeks had left off.

Their algebra was rhetorical (expressed in sentences rather than equations) and although they lacked algebraic symbols for operators they possessed something that was equally important, the concept of place value. They had also developed a set of number symbols (from 1 to 9) that was separate from their alphabet and could exploit a place-value system. Above all, they had conceived of a symbol (zero) to keep the number symbols in the correct columns. The Hindu symbol for 1 could only mean 1. If followed by a zero, it meant 10, if followed by two zeros it meant 100. The Indians called the symbol 0 *shunya*, "emptiness."

Our word "zero" derives ultimately from the Arabic *sifr*, which also gives us the word cipher. Zero is far from an intuitive concept and posed many difficulties to its early users—especially with division.

Bhaskara wrote after 500 years of its use in India that:

> *A quantity divided by zero becomes a fraction the denominator of which is zero. This fraction is termed an infinite quantity. In this quantity consisting of that which has zero for its divisor, there is no alteration, though many may be inserted or extracted; as no change takes place in the infinite and immutable God when worlds are created or destroyed, though numerous orders of beings are absorbed or put forth.*

Interest in classical history is apparent in a 13th-century Turkish manuscript that shows the 7th-century B.C.E. Athenian poet and statesman, Solon, engaged in a discussion with his students.

In 662 a Nestorian bishop named Severus Sebokht, who lived in a town on the Euphrates river, wrote:

> *I will omit all discussion of the science of the Indians ... of their subtle discoveries in astronomy, discoveries that are more ingenious than those of the Greeks and the Babylonians, and of their valuable methods of calculation which surpass description. I wish only to say that this computation is done by means of nine signs. If those who believe, because they speak Greek, that they have arrived at the limits of science, would read the Indian texts, they would be convinced, even if a little late in the day, that there are others who know something of value.*

It had taken the Hindus several centuries to fuse the three elements—number symbols, place value, and zero—but by the 7th century C.E. the system was well established. According to tradition the Indian numerals were introduced to the Islamic world by a Hindu scholar sometime around 770. Some fifty years later a Muslim mathematician named al-Khwarizmi published a treatise on arithmetic that explained the new place-value notation. His name was corrupted into the word "algorithm," which describes the step-by-step procedure of solving a mathematical problem. The title of his other major work, *Hisab al-Jabr wa-al Muqabalah* (*Book of Calculation of Restoration and Reduction*) gave rise to the word "algebra."

Persian Scholars

An 11th-century Persian geographer and physicist named al-Biruni completed the exposition of the new numeral system with a commentary on al-Khwarizmi that he produced after traveling to India. Their work was followed by the work of a perhaps even greater mathematician, a man better known in later times as a poet, Omar Khayyam.

Meanwhile in Europe, perhaps as a result of the infiltration of the Hindu numerals into Arab Spain, an improvement was made to the abacus that made its operation much easier. While the source of this new abacus is a matter of dispute, there is no question that division and multiplication could be performed with much greater ease and precision. The improvement, however, did not extend beyond the world of intellectual curiosities.

OMAR KHAYYAM

OMAR KHAYYAM (1048–1131)

A literal translation of the name al-Khayyam means "tent maker" and this may have been the trade of Omar's father Ibrahim. Khayyam played on the meaning of his own name when he wrote the following:

> *Khayyam, who stitched the tents of science,*
> *Has fallen in grief's furnace and been suddenly burned,*
> *The shears of Fate have cut the tent ropes of his life,*
> *And the broker of Hope has sold him for nothing!*

The Seljuk Turks were tribes that invaded southwestern Asia in the 11th century and eventually founded an empire that included Mesopotamia, Syria, Palestine, and most of Iran. Khayyam grew up in this unstable military empire.

He described the difficulties for men of learning during this period in the introduction to his *Treatise on Demonstration of Problems of Algebra*:

> *I was unable to devote myself to the learning of this algebra and the continued concentration upon it, because of obstacles in the vagaries of time which hindered me; for we have been deprived of all the people of knowledge save for a group, small in number, with many troubles, whose concern in life is to snatch the opportunity, when time is asleep, to devote themselves meanwhile to the investigation and perfection of a science; for the majority of people who imitate philosophers confuse the true with the false, and they do nothing but deceive and pretend knowledge, and they do not use what they know of the sciences except for base and material purposes; and if they see a certain person seeking for the right and preferring the truth, doing his best to refute the false and untrue and leaving aside hypocrisy and deceit, they make a fool of him and mock him.*

However, Khayyam was an outstanding mathematician and astronomer and, despite the difficulties he described, he authored several books during the quite extended periods of peace.

Of his many accomplishments, one that shows his technical abilities is his remarkably accurate measurement of the length of the year as 365.24219858156 days. We now know that the length of the year changes in the sixth decimal place over a person's lifetime, so that the length of the year at the end of the 19th century was 365.242196 days and today it is 365.242190 days.

When Khayyam was about fifty he came under attack from orthodox Muslims who felt that his questioning mind did not conform to the faith. He wrote in his famous poem the *Rubaiyat*:

> *Indeed, the Idols I have loved so long*
> *Have done my Credit in Men's Eye much Wrong:*
> *Have drown'd my Honor in a shallow Cup,*
> *And sold my Reputation for a Song.*

When calculations are made with a mechanical device, such as an abacus, there is no need to set down the operations or record their sequence. With written computations, it becomes a necessity. In the 16th century, students would turn their calculations into drawings, such as this one that resembles a ship. This method had been used in the 9th century by the Arab mathematician al-Khwarizmi.

During the reigns of Alfonso VI and VII of Castile, Toledo became a center where scholars from all over Europe gathered to glean scientific knowledge from the Arabs. The works of al-Khwarizmi had been translated and were studied there, but for some reason neither the numerals nor the place-value system reached beyond the confines of a small audience of learned men. Neither students nor businessmen knew anything about them.

Fibonacci's Grasp on the Matter

This was the situation when Leonardo arrived in Bugia at the end of the 12th century. Although he does not say exactly where or under what conditions he came into contact with the Indian notation, it is clear that he quickly grasped its significance. For business transactions the Indian system had obvious advantages over the abacus. It may have been even more exciting to the young mathematician to feel that the closed doors of calculation had been thrown open, for Leonardo clearly intuited the even greater possibilities that lay ahead.

We don't know how long Fibonacci stayed in Bugia, but we know he traveled throughout the Mediterranean world to study under the leading Arab mathematicians of the time, staying several years in Constantinople and visiting Egypt, Syria, Sicily, and Provence. He returned to Pisa around 1200 and immediately set to work writing a book about the wealth of new mathematics he had acquired, making significant contributions of his own.

In 1202, when he was twenty-seven, the book he had written was published—that is, copied by hand—in Pisa. Leonardo called the book *Liber abaci*, or the *Book of Calculation*. Its stated purpose was to introduce the Hindu place-value system to Europe and to explain the use of the new numerals. He proposed this system not only to elite scholars but to common people engaged in the world of commerce. The written procedures of calculation (as opposed to calculating with an abacus) were generally known in Italy at that time as *abaco*. So, while derived from the word abacus, the word *abaci* refers to calculation without the abacus.

The nine Indian figures are: 9 8 7 6 5 4 3 2 1. With these nine figures, and with the sign 0 ... any number may be written as is demonstrated below.
FIBONACCI'S OPENING SENTENCE IN *LIBER ABACI*

Fibonacci continued to add information to the book and in 1228 issued a revised edition:

> *In this rectification I added certain necessities, and I deleted certain superfluities...*

and it is the revised edition of 1228 that has come down to us. The first seven chapters of *Liber abaci* deal with the numerals themselves, what they are and how to use them both as whole numbers and as fractions. The second section contains techniques and applications of these techniques to practical problems of commercial bookkeeping—problems such as the conversion of weights and measures, the calculation of interest, and how to convert between the various currencies in use in Mediterranean countries. The rest of the book is devoted to the mathematics of series and proportion, how to solve problems, the extraction of roots, and geometry and algebra. *Liber abaci* was enthusiastically received throughout Europe and had a profound impact on European thought, although it was the practical applications rather than the abstract theorems that made Fibonacci famous to his contemporaries.

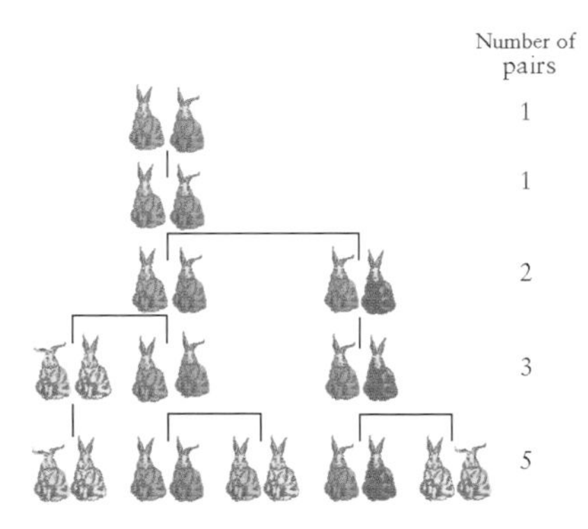

Each pair of rabbits continues to breed one pair per month, once they start. With each new month every pair that is breeding continues to breed. The number of pairs at the end of any given month is the sum of the two numbers preceding it.

The Fibonacci Sequence

A problem that appears in the third section of *Liber abaci* led to the discovery for which Fibonacci is best remembered today—the Fibonacci numbers and the Fibonacci sequence:

> *A certain man put a pair of rabbits in a place surrounded on all sides by a wall. How many pairs of rabbits can be produced from that pair in a year if it is supposed that every month each pair begets a new pair, which from the second month on becomes productive?*

The resulting sequence is 1, 1, 2, 3, 5, 8, 13, 21, 34, 55 ... (Fibonacci omitted the first term of the complete sequence in *Liber abaci*). This sequence, in which each number is the sum of the two preceding numbers, completely surprised future mathematicians when they discovered its relationship to the Golden Mean, our Divine Proportion. The ratio of any neighboring numbers in the series approaches the Golden Mean. Furthermore, as if by some uncanny inner dynamism, the numbers in the series can be expressed in a number of interesting mathematical equations and other very surprising relationships. Over the many years that have followed the expression of this series in the famous rabbit problem, Fibonacci numbers have shown up in the unfolding leaves and seed patterns in plants and, curiously, in the family trees of bees. Its numbers are associated with regenerative spirals that occur throughout the human body and that can be seen displayed in the growing shapes of shells and galaxies.

The Works of Fibonacci

Fibonacci lived in the days before printing, so the only way to have a copy of the book was to have another hand-written copy made. Of the many books he wrote we still have copies of *Liber abaci* (1202), *Practica geometriae* (1220), *Flos* (1225), and *Liber quadratorum* (1225). Given that relatively few copies would ever have been produced, we are fortunate to have these, and we know there are other texts that, unfortunately, were lost. His book on commercial arithmetic, *Di minor guisa,* was lost, as was his commentary on Book X of Euclid's *Elements* that contained a treatment of irrational numbers.

Arithmetica by Gregor Reisch (1467–1525)

Boethius and Pythagoras are engaged in a mathematical competition. Pythagoras is using an abacus, while Boethius uses numerals from India. Boethius looks very proud, for he has already finished his calculation while poor Pythagoras still tries to find the solution.

During Fibonacci's lifetime, Frederick II was crowned king of Germany and then crowned Holy Roman emperor by the pope in 1220. He became aware of Fibonacci's work through the scholars at his court who had corresponded with Fibonacci since his return to Pisa in around 1200. These scholars included Michael Scotus, who was the court astrologer (to whom Fibonacci dedicated *Liber abaci*), Theodorus Physicus, the court philosopher, and Dominicus Hispanus, who suggested to Frederick that he meet Fibonacci when Frederick's court met in Pisa around 1225.

Johannes of Palermo, another member of Frederick II's court, presented a number of problems as challenges to the great mathematician. Fibonacci gives the solutions to three of them in *Flos.* In one of these solutions, Fibonacci gives an accurate approximation to a root of $10x + 2x^2 + x^3 = 20$. Palermo had taken this problem from Omar Khayyam's algebra book, where it is solved

Fibonacci Numbers

THESE ARE THE FIRST TWENTY-FIVE FIBONACCI NUMBERS.

Every 3rd Fibonacci number is a multiple of 2.

Every 4th Fibonacci number is a multiple of 3.

Every 5th Fibonacci number is a multiple of 5.

Every 6th Fibonacci number is a multiple of 8.

1
1
2
3
5
8
13
21
34
55
89
144
233
377
610
987
1597
2584
4181

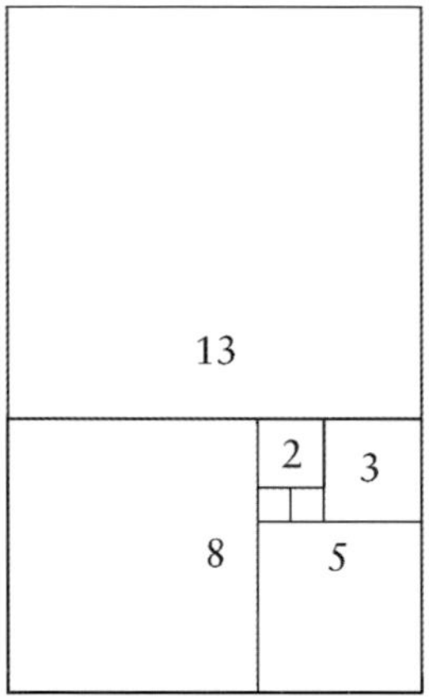

13
2
3
8
5

THE GOLDEN RECTANGLE or Rectangle of Divine Proportion can be drawn by nesting squares of Fibonacci numbers next to each other.

We can see that each rectangle is made up of all the earlier squares, all of which have sides that are Fibonacci numbers in length. The diagram shows that the area of each rectangle is the product of the sides of the last square added to the next one in the series.

$1^2 + 1^2 = 1 \text{x} 2$

$1^2 + 1^2 + 2^2 = 2 \text{x} 3$

$1^2 + 1^2 + 2^2 + 3^2 = 3 \text{x} 5$

$1^2 + 1^2 + 2^2 + 3^2 + 5^2 = 5 \text{x} 8$

$1^2 + 1^2 + 2^2 + 3^2 + 5^2 + 8^2 = 8 \text{x} 13$

FIBONACCI NUMBERS AND DIVINE PROPORTION

If we look at the numbers in the Fibonacci sequence and find the ratios between successive numbers, we will see that they approach the Divine Proportion. Remember that

$$\Phi = \frac{(1+\sqrt{5})}{2} = 1.61803398874989484820458683....$$

0	
1	
1/1	1.000000000000000
2/1	2.000000000000000
3/2	1.500000000000000
5/3	1.666666666666667
8/5	1.600000000000000
13/8	1.625000000000000
21/13	1.615384615384615
34/21	1.619047619047619
55/34	1.617647058823529
89/55	1.618181818181818
144/89	1.617977528089888
233/144	1.618055555555556
377/233	1.618025751072961
610/377	1.618037135278515
987/610	1.618032786885246
1,597/987	1.618034447821682
2,584/1,597	1.618033813400125
4,181/2,587	1.618034055727554
6,765/4,181	1.618033963166707
10,946/6,765	1.618033998521803
17,711/10,946	1.618033985017358
28,657/17,711	1.618033990175597
46,368/28,657	1.618033988205325
75,025/46,368	1.618033988957902
121,393/75,025	1.618033988670443
196,418/121,393	1.618033988780243
317,811/196,478	1.618033988738303
514,229/317,811	1.618033988754323
832,040/514,229	1.618033988748204
1,346,269/832,040	1.618033988750541

Miniature from a 15th-century manuscript of Sacrobosco, a 13th-century English mathematician who played an important part in the dissemination of the new numerals.

by means of the intersection of a circle and a hyperbola. Fibonacci proves that the root of the equation is neither an integer nor a fraction, nor the square root of a fraction. He says:

> *And because it was not possible to solve this equation in any other of the above ways, I worked to reduce the solution to an approximation.*

Without explaining his methods, Fibonacci gives the approximate solution in sexagesimal notation as 1.22.7.42.33.4.40 (this is written to base 60, so it is $1 + 22/60 + 7/60^2 + 42/60^3 + \ldots$). In base 10 numbers this results in the decimal 1.3688081075, which is correct to nine decimal places and is thought to be a remarkable achievement.

Liber quadratorum, written in 1225, is Fibonacci's most impressive piece of work, although not the work for which he is most famous. The book's name means *The Book of Squares* and it is a number theory book, which, among other things, examines methods to find Pythogorean triples (integers that satisfy the equation $a^2 + b^2 = c^2$, for example 3, 4, 5).

> *Thus when I wish to find two square numbers whose addition produces a square number, I take any odd square number as one of the two square numbers and I find the other square number by the addition of all the odd numbers from unity up to but excluding the odd square number. For example, I take 9 as one of the two squares mentioned; the remaining square will be obtained by the addition of all the odd numbers below 9, namely 1, 3, 5, 7, whose sum is 16, a square number, which when added to 9 gives 25, a square number.*

Fibonacci Numbers and Probability

Mathematicians have long been intrigued with dice and probability. There is a story told that Blaise Pascal and Pierre de Fermat were sitting in a Paris café discussing different problems and decided to play a game of chance. They began flipping a coin for points and money. Before they finished Fermat was called away. This led them to consider an interesting problem:

Two equally skilled players are interrupted while playing a game of chance for a certain amount of money. Given the score of the game at that point, how should the stakes be divided?

In 1654 the two men worked out their theory of probability in their correspondences and launched the study of probability as a new branch of mathematics. Pascal spent much time studying the triangle, which we now call Pascal's triangle; it is the basis for some particular properties of probability.

Although Pascal was not aware of it, the Fibonacci numbers appear in the triangle.

In 1642 Pascal invented the first adding machine, and in 1673 Gottfried Wilheim von Leibnitz invented one that could also multiply and divide. One hundred and fifty years earlier Leonardo da Vinci had conceived the idea and made drawings of a mechanical counting device.

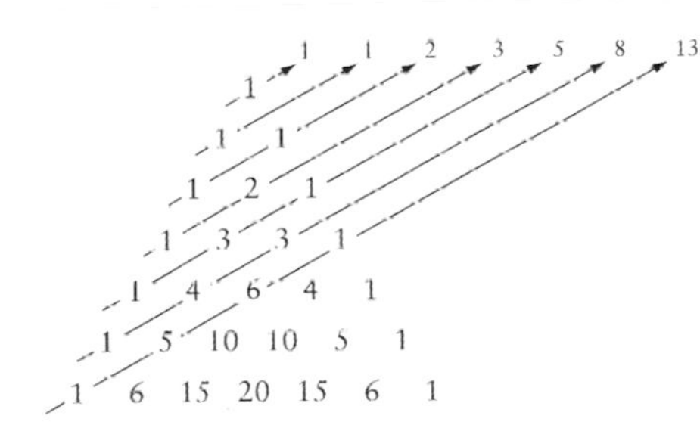

And regarding square numbers:

> *I thought about the origin of all square numbers and discovered that they arose from the regular ascent of odd numbers. For unity is a square and from it is produced the first square, namely 1; adding 3 to this makes the second square, namely 4, whose root is 2; if to this sum is added a third odd number, namely 5, the third square will be produced, namely 9, whose root is 3; and so the sequence and series of square numbers always rise through the regular addition of odd numbers.*

This means that for any square number, n, the next square number can be found using $n^2 + (2n+1) = (n+1)^2$.

Fibonacci's love for pure mathematics—the theories that underlie all applications—was extraordinary for his time and was coupled with a real zeal to see the practical application of the new numerals in the world of commerce. Although he was clearly well known to many people in his own time, his fame was postponed for many generations. After 1228 there is only one known document that refers to Fibonacci—a decree made by the Republic of Pisa in 1240 in which an annual honorarium is awarded to:

> *... the serious and learned Master Leonardo Bigollo ....*

There is no information about how long Fibonacci lived or how his life ended. Two streets in Italy—a street in Pisa called the Lungarno Fibonacci and one in Florence, the Via Fibonacci—are the sole public tributes to his contribution.

The Hindu numerals, his great bequest to Western European civilization, at first encountered widespread opposition. People, naturally resistant to change and already conversant with the Roman symbols, suspected that the new numbers could be too easily manipulated. Conservatives charged that the numbers were difficult to learn. However, by the 15th century they were displacing both the Roman numerals and the abacus in commerce. Coins were struck using the new numerical values, and Hindu numerals made their way onto calendars while the Roman symbols passed into the secondary status that they have occupied ever since.

Allegory of Arithmetic

In this 16th-century tapestry Lady Arithmetic teaches the art of calculation to some youths. Many of the great artists of the Renaissance were well versed in the manipulation of numbers.

Mathematical application of the new numerals spread with the invention of the printing press and the real awakening that Fibonacci foresaw began. European mathematicians took up algebra again, probably due to the greater ease of calculation, and the 16th and 17th centuries saw the extraordinary progress of Cavalieri, Fermat, Pascal, Descartes, Kepler, and Napier, followed in the 18th century by the works of Newton and Leibnitz.

There is no doubt that Fibonacci was far ahead of his time. Pure mathematics (his real passion) did not take off until three hundred years after his death. As a result, his historic contribution was overlooked and few of his works have been translated.

> *Philosophy is written in this grand book—I mean the universe—which stands continually open to our gaze, but it cannot be understood unless one first learns to comprehend the language and interpret the characters in which it is written. It is written in the language of mathematics, and its characters are triangles, circles, and other geometrical figures, without which it is humanly impossible to understand a single word of it; without these, one is wandering about in a dark labyrinth.*
>
> GALILEO GALILEI (1564–1642)

Chapter 4

Divine Proportion in Architecture, Art, and Music

Once organic character is achieved in the work of Art, that work is forever. Like sun, moon, and stars, great trees, flowers, and grass it is and stays on while and wherever man is.

FRANK LLOYD WRIGHT

Art is an experience of balance, of the relationship of its parts to the whole. Perceiving it as anything else is missing its most fundamental component. A fine painting, a piece of sculpture, a work of architecture, music, prose, or poetry is organized and gracefully balanced around a hidden sense of proportion.

The particular harmony of the Divine Proportion was noticeably built into the Gothic cathedrals of Europe and Le Corbusier's modular designs in contemporary architecture. The same proportion appears in the paintings of Leonardo da Vinci, Albrecht Dürer, and Georges Seurat, and in the sculptures of Phidias and Michelangelo, and lies at the heart of all music. It is the same proportion that is symbolically carried into many works of art in the use of mystic spirals, triangles, pentagrams, and Golden

Jimson Weed by Georgia O'Keefe (1887–1986)

I had to create an equivalent for what I felt about what I was looking at—not copy it.

GEORGIA O'KEEFE

OPPOSITE: *The Last Supper* by Leonardo da Vinci (1452–1519)

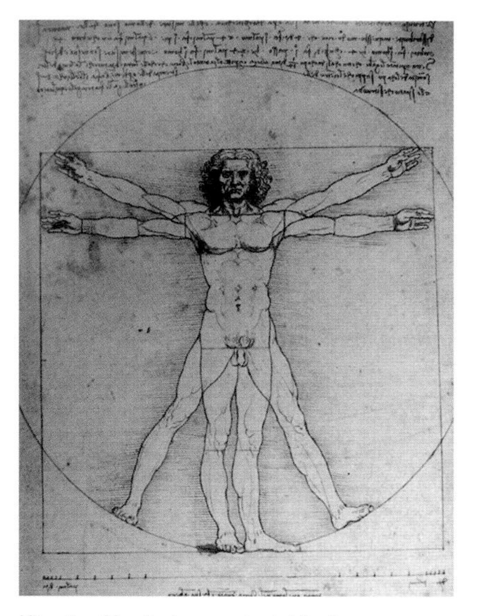

Vitruvian Man by Leonardo da Vinci

PROPORTION IN THE BODY

Leonardo da Vinci's famous drawing, *Vitruvian Man,* appeared in the 1509 book *Divina proportione* by Luca Pacioli.

Leonardo wrote extensively in his notebooks about the proportions of the human body. He determined measurements and proportions for all its parts, basing his studies on numerous observations and measurements. He referred in the notebooks to the works of Vitruvius:

Vitruvius, the architect, says in his works on architecture that the measurements of the human body are distributed by Nature as follows: ...If you open your legs so much as to decrease your height by 1/14 and spread and raise your arms till your middle fingers touch the level at the top of your head you must know that the center of the outspread limbs will lie in the navel and the space between the legs will be an equilateral triangle.

Rectangles. This proportion is used in both the most subtle and the most obvious ways to impart a sense of structural harmony, balance, and divinity. It can be seen in the overall dimensions of a work or in the divisions or parts of the whole. It can be seen in the content, or it can be felt. It can be strictly adhered to or slightly departed from; and it can spring from a conscious or an unconscious impulse in the artist.

Heraclitus (540–480 B.C.E.), one of the most significant mystics of ancient Greece, once said, "Man is the measure of all things." Man or woman, we are the very balancing of the Divine Proportion. We are represented by the segment on Euclid's wonderful line that is proportionate to the whole in the same way it is to the balance of the line, and whatever we describe is a reflection of our relationship to the whole. Blessed with an innate sense of this proportion, our highest expression becomes art. When we deny the feeling of proportion within ourselves, we produce nothing of enduring significance.

ORIGIN OF THE VITRUVIAN MAN

The relationship of the Divine Proportion and the human body was first referred to in the pentagram described by the pentad in the Pythagorean science of numbers. This relationship was further defined by Marcus Vitruvius Pollio (Vitruvius), a Roman writer, architect and engineer. Around 27 B.C.E. he wrote a book called *De Architectura*, known today as *The Ten Books of Architecture*. This was a treatise on architecture that covered an enormous range of subjects: building materials, temple construction, public buildings (theatres, baths), private buildings, floors and stucco decoration, hydraulics, clocks, and civil and military engines. This book is based primarily on Greek models, though

MICHELANGELO AND LE CORBUSIER

Statue of *David* by Michelangelo

Many artists of the Renaissance used the Divine Proportion as a design element in their work. Michelangelo's sculpture *David* conforms to it.

MICHELANGELO

Michelagniolo Buonarroti (1475–1564) became an artist at the age of thirteen against his father's wishes. Originally apprenticed as a painter, he rapidly achieved fame as a sculptor, and later in life as a poet and architect. The range of his talent is highlighted by his four greatest works: *Pietà* in St. Peter's in Rome, the *Last Judgement* on the walls of the Sistine Chapel in the Vatican, the dome of St. Peter's Basilica in Rome, and the statue of *David* in Florence. *David* was carved from a single piece of white Carrara marble that was rejected by other sculptors as being too shallow. It represents the moment when the youth is about to slay Goliath, a moment that is filled with political and cultural symbolism. For the Florentines this statue represented the ideal of the self-sufficient republic ready to withstand the pressure of its neighbors. This single work established Michelangelo as the greatest sculptor in Italy.

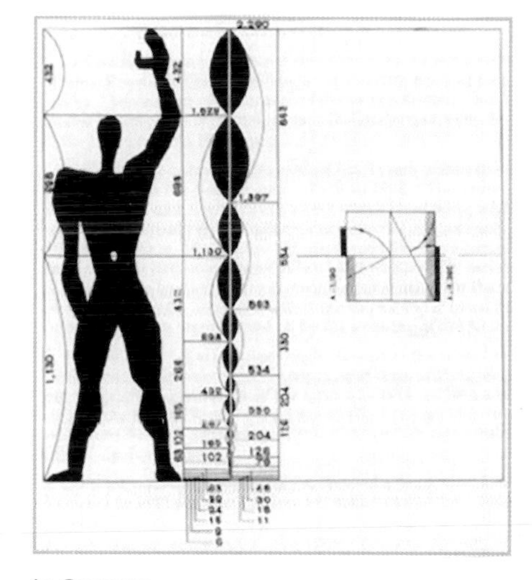

LE CORBUSIER

Le Corbusier created a system of proportions he called Modulor in order to avoid converting feet or inches into the metric system. He began by using just a few measurements on his construction sites that corresponded to proportions of a human body. This system was further elaborated using the Golden Section and the Fibonacci series of numbers, and application of these proportions can be found in many of his buildings.

Le Corbusier (1887–1965) was born Charles-Edouard Jeanneret in Switzerland. When he was twenty-nine, he went to Paris and adopted his maternal grandfather's name, Le Corbusier, as his pseudonym. He believed that architecture had lost its way and spent two months studying the Parthenon and other ancient Greek buildings. "The Parthenon is certainly one of the purest works of art that man ever made," he said.

The ruined temple of Sunion on Mt. Olympus

Roman architecture soon departed from Greek models in order to serve the new needs of the growing Roman empire. Vitruvius wished to preserve the classical tradition in the design of temples and public buildings, and a few of his prefaces contain pessimistic remarks about the architecture of his time. Throughout the Renaissance his book was the chief authority on ancient classical architecture. In Book Three he discusses temples.

The Planning of Temples

> *1. The design of a temple depends upon symmetry, the principles of which must be most carefully observed by the architect. They are due to proportion, in Greek* analogia. *Proportion is a correspondence among the measures of the members of an entire work, and of the whole to a certain part selected as standard. Without symmetry and proportion there can be no principles in the design of any temple; that is, if there is no precise relation between its members, as in the case of those of a well-shaped man.*
>
> *2. For the human body is so designed by nature that the face, from the chin to the top of the forehead and the lowest roots of the hair, is a tenth part of the whole height; the open hand from the wrist to the tip of the middle finger is just the same; the head from the chin to the crown, an eighth; and with the neck and shoulder from the top of the breast to the lowest roots of the hair is a sixth; from the middle of the breast to the summit of the crown is a fourth. If we take the height of the face itself, the distance from the bottom of the chin to the under side of the nostrils is one third of it; the nose from the under side of the nostrils to a line between the eyebrows is the same; from there to the lowest roots of the hair is also a third, comprising the forehead. The length of the foot is one*

sixth of the height of the body; of the forearm, one fourth; and the breadth of the breast is also one fourth. The other members, too, have their own symmetrical proportions, and it was by employing them that the famous painters and sculptors of antiquity attained to great and endless renown.

3. Similarly, in the members of a temple there ought to be the greatest harmony in the symmetrical relations of the different parts to the general magnitude of the whole. Then again, in the human body the central point is naturally the navel. For if a man be placed flat on his back, with his hands and feet extended, and a pair of compasses centred at his navel, the fingers and toes of his two hands and feet will touch the circumference of a circle described therefrom. And just as the human body yields a circular outline, so too a square figure may be found from it. For if we measure the distance from the soles of the feet to the top of the head, and then apply that measure to the outstretched arms, the breadth will be found to be the same as the height, as in the case of plane surfaces which are perfectly square.

4. Therefore, since nature has designed the human body so that its members are duly proportioned to the frame as a whole, it appears that the ancients had good reason for their rule, that in perfect buildings the different members must be in exact symmetrical relations to the whole general scheme. Hence, while transmitting to us the proper arrangements for buildings of all kinds, they were particularly careful to do so in the case of temples of the gods, buildings in which merits and faults usually last forever.

Chartres Cathedral in 1696 in an engraving by Pierre Ganière

This passage inspired Leonardo da Vinci to draw the now famous *Vitruvian Man* in which the proportions Vitruvius

Artemis from the freize of the Parthenon

speaks of were so brilliantly translated in one of his journals. The two superimposed positions of the naked man with his arms outstretched describe a circle centered on his phallus and a square centered on the true center of gravity, his navel. The legs are spread to 60 degrees and the knees, penis, and nipples divide the height into equal quarters.

Φ, Phidias and the Divine Proportion

Phidias, also spelled Pheidias (ca. 490–430 B.C.E.), from whose name we have the symbol Φ (Phi), is believed to be one of the first sculptors to intentionally put the Divine Proportion to use. He is universally regarded as the greatest Greek sculptor.

While none of Phidias' original statues have survived, numerous Roman copies of his work in varying degrees of supposed fidelity exist. His fame has been unquestionably secured through the descriptions of ancient writers and their recognition of the enormous influence he had on sculpture of his time.

Phidias lived during the Persian Wars of 492–449 B.C.E—a series of wars fought on the Greek mainland against invading Persians. By the time the war was over and the Persians defeated, they had twice invaded Athens and sacked the city, forcing the Athenians to flee. The Athenians returned to find their homes and temples destroyed. Pericles, a statesman who had recently risen to power, persuaded the Athenians to undertake a great building program, and in 447 B.C.E. he appointed Phidias to oversee the work. Phidias sculpted, supervised, and maybe even designed the overall sculptural decoration of the Parthenon's rebuilding. The Greeks said that Phidias alone had seen the exact image of the Gods and that this was revealed in the statues of Athena and of Zeus.

THE ACROPOLIS

THE ACROPOLIS OR "SACRED ROCK" of Athens was built on a site that had been sacred from as early as 1300 B.C.E. (the end of the Bronze Age). The first habitation remains date from the Neolithic period. Over the centuries, the hill had been continuously used as a center for religious ceremonies.

In the construction of the Acropolis three important temples were erected on the ruins of earlier ones: the Parthenon, the Erechtheion, and the Temple of Nike—dedicated to Athena Parthenos, Athena Polias, and Athena-Apteros Nike, respectively. The Propylaia, the monumental entrance to the sacred area was also constructed in the same period.

The Acropolis was rebuilt in the wake of Athens' final victory over the Persians, and those who ascended the Acropolis, after having passed through the Propylaia, had their view and way barred by the ruins of the Old Temple of Athena. This temple had been sacked, burned, and wrecked by the Persians when in 480 B.C.E. the Athenians made the extreme sacrifice of abandoning their homes and temples to the enemy. The ruins of this temple were used to justify Athens' claim to leadership of the Greek world and to justify transforming that leadership into imperial domination.

Because of this, the temple that was intended to replace the destroyed temple was not erected on top of the old one, as might have been expected, but to the south of it.

THE PARTHENON

The Parthenon was built between 447 and 438 B.C.E. and dedicated to Athena Parthenos, the patron Goddess of Athens.

The temple has eight columns on each of the narrow sides and seventeen columns on each of the long ones. Phidias' famous statue of Athena stood in the central part of the temple. The frieze on the temple's walls depicts the Procession of the Panathenaea, the most formal religious festival of ancient Athens. The scene runs along all the four sides of the building and includes the figures of gods, beasts, and some 360 humans.

THE PROPYLAIA

The entrance to the Parthenon was the Propylaia—two enormous wings oriented to what was then the most important area of the Acropolis, a point just north of the North Portico where the olive of Athena grew inside the ancient Erechtheion. After the Persians reduced the temple to rubble and the olive tree to a smoldering stump, the Erechtheion was rebuilt so that its western wall formed a backdrop that allowed visitors to admire the tree's new shoots. Earlier, there had been a road that went from the temple to the tree.

PHIDIAS' WORK

ATHENA PARTHENOS
This very heavy Roman copy is the only evidence of the glorious appearance of the gold and ivory Athena Parthenos made by Phidias for the Parthenon.

The Goddess Athena was born to Zeus and his first wife, Metos (Wisdom). The myth of her birth tells us that Zeus had been advised that any children he had by Metos would be very powerful and eventually dethrone him, so when Metos was about to give birth, Zeus swallowed her. However, he was soon tortured by a severe headache. As a cure, Hephaestus, the God of fire, split his skull with an ax and from the wound sprang a fully armed Athena.

As Goddess, Athena served many roles. Stones falling from the sky were among the earliest representations of her, seeming to suggest her origin as a lunar Goddess or some association with meteorites. Most well known as the patron Goddess of the city of Athens, she was charged with protecting many cities, and so there are temples to her throughout Greece. She also became a Goddess of wisdom, and the owl was often adopted as her symbol.

THE RUINS OF ZEUS' TEMPLE AT OLYMPIA (ca. 466–456 B.C.E.) Inside the temple stood Phidias' world-famous statue of Zeus.

A coin from Alexandria (ca. 315–310 B.C.E.) shows Athena striding into battle carrying her spear and shield in her hands.

Minted at the time of the Persian wars, this coin shows the owl, sacred to Athena, with the initials of the great city that twice prevailed over the Persians.

In his Hymn, the poet Kleanthes (4th/3rd century B.C.E.) writes of Zeus:

> *This entire cosmos which moves round the Earth follows where you lead it, and is willingly mastered by you… Nothing happens in the world without you, God, neither in the divine air of the sky, nor in the sea, except for that which evil men do in their folly.*

His beautiful creation of Athena Parthenos stood thirty feet tall inside the Parthenon. She held a figurine of Nike (Goddess of victory) in her right hand and a spear in her left, and had a decorated shield and a serpent by her side. The original statue was made of gold and ivory. Several copies—both Greek and Roman—have been identified.

Zeus enthroned with an eagle and a scepter on a Macedonian coin

The last years of Phidias' life are somewhat of a mystery. Pericles' enemies accused Phidias of stealing gold from the statue of the Athena Parthenos, but he was able to disprove the charge. They then accused him of impiety for including portraits of Pericles and himself on the shield of the Athena. For this he was thrown into prison and until recently it was thought that he died there a short time later. Now it is believed that he was exiled to Elis where he worked on the Olympian Zeus.

Head of Zeus on a Roman coin from the time of the emperor Hadrian (76–138 C.E.)

We have unfortunately lost all trace of the Zeus, except for small copies on coins that give us nothing more than a notion of the pose and the character of the head. Phidias had placed the God on a throne with a Nike in his right hand and a scepter in his left. Like Athena Parthenos, his flesh was of ivory and his drapery of gold. Ancient writers believed the statue stood a towering forty-three feet and it is now considered one of the Seven Wonders of the Ancient World.

In 1958 archaeologists found the workshop at Olympia where Phidias assembled the statue of Zeus. There were still some shards of ivory at the site and the base of a bronze drinking cup engraved "I belong to Phidias."

From the interior of the Parthenon looking east

Divine Proportions of the Parthenon

The Parthenon that housed Athena Parthenos is the most well-known building of the Acropolis, a temple compound that stands in the center of Athens on an outcrop of rock dominating the ancient city. There have been numerous attempts to determine the system of proportions used to design the Parthenon, on the assumption that this knowledge would reveal the secret of its beauty and of Greek architecture in general.

Vitruvius, in listing the writers on architecture that constitute the sources of his book, mentions that Ictinos, the architect of the Parthenon, explained the proportions of this temple in a book. Although this particular work is lost to us, we can presume on the basis of the importance of the book, that these proportions constituted a system that was of intellectual significance and appeal. Since the explanation of these proportions is lost, Ictinos speaks to us only through the stonework of his creation—stones that were carefully put together so that they would last for eternity.

The Parthenon has had many uses through the centuries, and it remained in good condition until it was hit by a Venetian bomb in 1687. A large enough portion still stands to convey Ictinos' message of respect for the goddess and the Athenian soul. The construction is so harmonious its beauty can still be felt today.

Based on modern calculations the Parthenon appears to be built on a square-root-of-5 rectangle, that is, a rectangle with its sides the length of the irrational $\sqrt{5}$. The front elevation is built as a Golden Rectangle. Because the top part is missing and the base is curved to counteract an optical illusion that level lines are bowed, this is only an approximate measure.

DIVINE PROPORTION AND THE PARTHENON

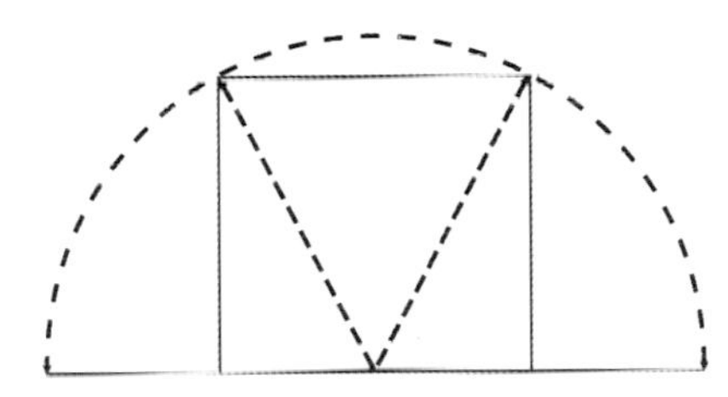

The square-root-of-5 rectangle is found when a unit square is placed in a circle as in the drawing above and a rectangle is drawn.

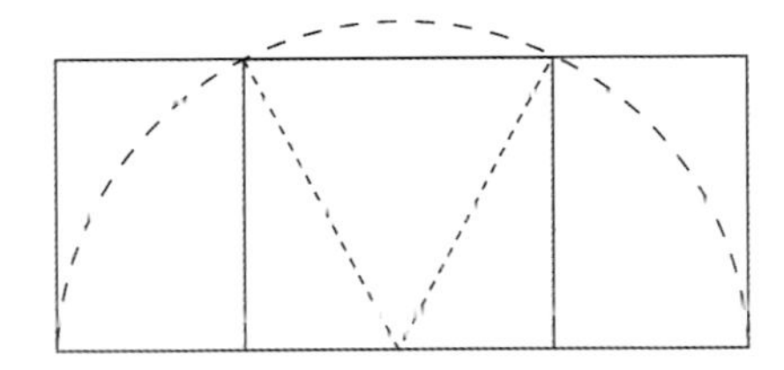

The small rectangles on either side of the square are Golden Rectangles, and when one of these Golden Rectangles is added to the center square another Golden Rectangle is formed. Both Golden Rectangles and the square between them form a square-root-of-5 rectangle.

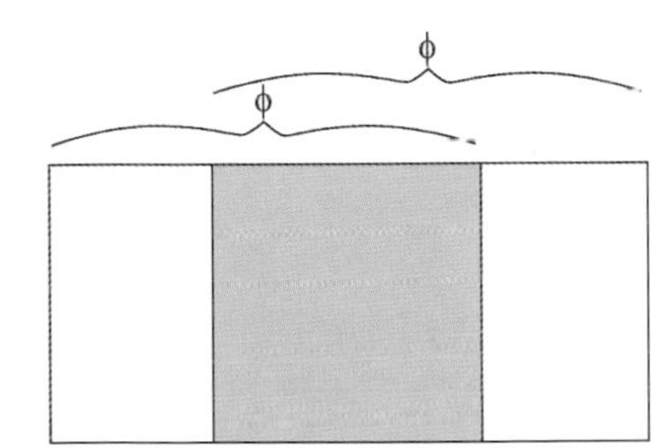

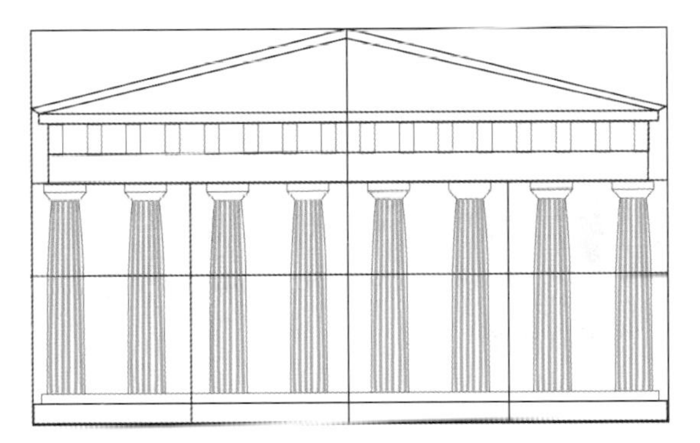

While its triangular pediment was still intact, the dimensions of the Parthenon fit into a Golden Rectangle. Its floor plan appears to be based on a square-root-of-5 rectangle.

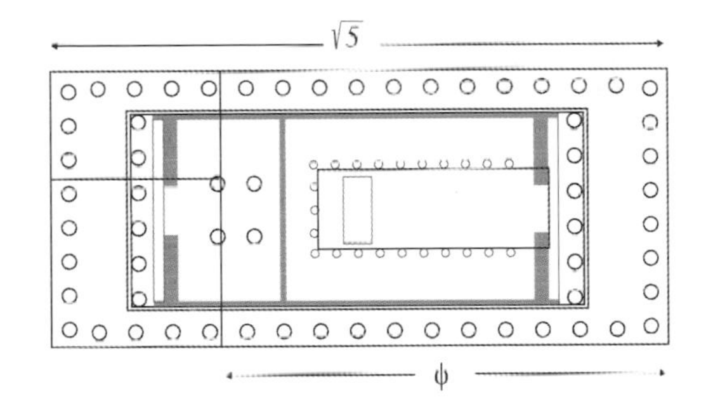

Notre Dame de Paris

The west face of the most famous of the Gothic cathedrals is rich in relationships defined by the Divine Proportion.

The original design for the towers of Notre Dame included steeples, but they were never added to the structure.

Medieval Cathedrals

The builders of medieval churches and cathedrals in a later era approached the design of their buildings in much the same way as the Greeks had approached theirs. Harmonic structure and grace was their aim. These magnificent churches and cathedrals adhere to a design of perfect and beautiful proportions, in the same way that the Parthenon had. Inside and out, these medieval buildings were intricate constructions based on the Golden Section and other rules of proportion.

Notre Dame de Paris (French for "Our Lady of Paris"), located on the eastern half of the Île de la Cité in Paris, was dedicated to Mary, the mother of Jesus, and is often known simply as Notre Dame. It is one of the most famous cathedrals of the Middle Ages and probably the most famous image in French Gothic art. Rising to 110 feet, it was the first cathedral to be built on a truly monumental scale and became a prototype for future French cathedrals. Its west façade is beautifully designed to the Divine Proportion.

On the spot where the cathedral now stands the Celts originally worshiped their gods, and the Romans later built a temple to Jupiter. This was followed by a Christian basilica and then a Romanesque church (the Cathedral of St. Etienne, founded by Childebert in 528). Some accounts claim that two churches existed on the site, one dedicated to the Virgin Mary and the other to St. Stephen. In 1163 Maurice de Sully, bishop of Paris, decided to build a new cathedral for the expanding population and to dedicate it to the Virgin Mary. Construction was not completed until roughly 180 years later. Built in an age of illiteracy, the cathedral retells the stories of the Bible in its portals, paintings, and stained glass.

Building of Gothic Cathedrals

Cathedral building required high-quality stones that could withstand the effects of time. Within three centuries, from 1050 to 1350, several million tons of stone were quarried in France in order to construct eighty cathedrals, five hundred large churches, and several tens of thousands of parish churches. More stone was cut in three centuries in France than in any period in the history of Egypt, even though the Great Pyramid is estimated to have about 2,300,000 stone blocks weighing from two to thirty tons each.

One of three 13th century rose windows of Chartres Cathedral

Gothic architecture was characterized by a reduction in the surface area of the masonry walls of the building. Stained-glass windows gradually replaced the fresco, while serving the same purpose of transmitting a message or a lesson to those who could not read.

Tradition tells us that when the first outlines of the cathedral of Cologne began to emerge, the Devil appeared to the architect, Gerhard, and threatened to stop him from completing the building by causing a canal to emerge in the middle of the foundations. Gerhard did not hesitate to wager his soul, for he alone knew the secret of how the canals were built and, because of this, was convinced that the Devil would be unable to carry out his threat.

When he went home, Gerhard is said to have told the secret to his wife, but the Devil elicited it from her by some cunning means. A short time later, when Master Gerhard saw water flowing from the foundation, he thought that he had lost his soul. He jumped down from the scaffolding into the void, and Satan pursued him in the form of a dog. The legend ends by stressing that no builder was able to finish the cathedral after that; and in fact the towers remained unfinished for centuries (until 1880).

Villard de Honnecourt worked for the Cistercian Order as an architect between 1225 and 1250 C.E. Although little is known of his responsibilities or his role in the construction of specific churches he is remembered for his collection of drawings. Apparently he traveled from site to site capturing with his pen and pencil the great monuments and their builders as he saw them.

Saint Francis Preaching to the Birds by Giotto (1267–1337)

Giotto is regarded as a founder of modern Western painting because his work broke free from the stylizations of Byzantine art and introduced a convincing sense of three-dimensional space. His work stunned the poet Dante and other contemporaries, for he filled his pictures with human emotions, heralding a return to the poignant realism of Classical Greek art.

Copernicus's view of the solar system with the Sun at the center

The Renaissance

At the end of the 12th century Italy saw the emergence of what is known as Humanism, a philosophy that drew inspiration from St. Francis of Assisi who had gone out among the poor, praising the spiritual value of nature and rejecting the formal rigidity of prevailing Christian theology. Inquiring minds of men like Dante and Petrarch and the work of artists like Giotto, with their concerns for the inner experience, paved the way for what was to emerge as the Renaissance or "rebirth" of classical learning. The fall of Constantinople in 1453 provided Humanism with a major boost, for many Eastern scholars fled to Italy, bringing with them important books and manuscripts and the tradition of Greek scholarship.

Humanism had several significant features. First, it took human nature as its subject and emphasized the dignity of the individual. In place of the medieval ideal of a life of penance as the highest and noblest form of human activity, the Humanists looked forward to a rebirth of a lost human spirit and wisdom. Its effect was to help people break free from the mental strictures imposed by religious orthodoxy and to inspire free inquiry, criticism, and a new confidence in the possibilities of human thought and works of art.

The Renaissance—literally "rebirth"—refers to the period of time in Europe that saw the decline of the feudal system, the discovery of new continents, the replacement of the Ptolemaic system of astronomy with the Copernican one, and the invention of printing (roughly 1450–1600). To the scholars and thinkers of the day, however, it was primarily a time of the revival of classical learning and wisdom after a very long period of cultural decline and stagnation.

The spirit of the Renaissance is most readily found in painting. Art was seen as a branch of knowledge, valuable in its own right and capable of providing images of God and creation as well as insights into the place of humans in the universe. In the hands of men like Leonardo da Vinci it became a science and was a means of exploring nature and recording discoveries. In the works of painters such as Masaccio, Fra Angelico, Botticelli, Perugino, Piero della Francesca, Raphael, and Titian; sculptors such as Pisano, Donatello, Verrocchio, Ghiberti, and Michelangelo; and architects such as Alberti, Brunelleschi, and Palladio, human dignity found expression in the arts as well as in the ancient mathematical principles of balance, harmony, and perspective.

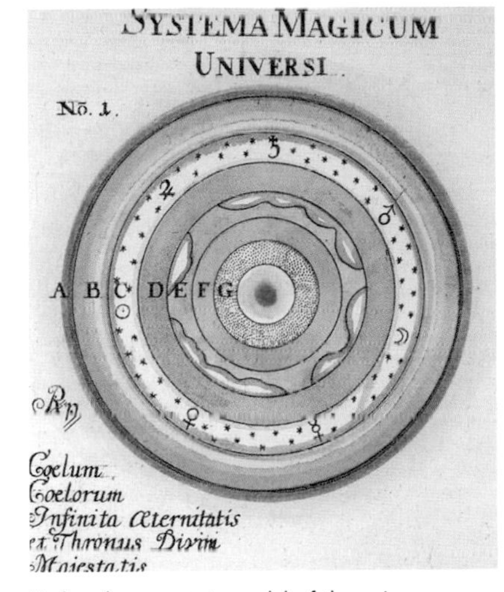

Ptolemy's geocentric model of the universe

LUCA PACIOLI AND *DIVINA PROPORTIONE*

Luca Pacioli (1445–1514), sometimes called "Paciolo," was one of the great men of the Renaissance, though one of the least well known. He was born in Sansepulcro, Tuscany, to a poor family, and his future seemed very unpromising. Having joined a Franciscan monastery in Sansepulcro he was apprenticed to a local businessman, but the young Pacioli, who loved mathematics, impetuously abandoned his apprenticeship to work as a mathematics scholar and befriended the artist Piero della Francesca, one of the first writers to explore and write about perspective. Francesca and Pacioli journeyed over the Appenines to Venice, where Francesca gave Pacioli access to the library of Frederico, the Count of Urbino. The collection of four thousand books allowed Pacioli to further his knowledge of mathematics and to write his first book on arithmetic. He left Venice and traveled to Rome where he spent several months living in the house of Leone Battista Alberti, who was a secretary in the Papal Chancery. Alberti was able to provide Pacioli with good religious

Luca Pacioli is the central figure in this painting by Jacopo de Barbieri (ca.1440–1515), which epitomizes the deep connection that existed between art and mathematics in the Renaissance.

THE GOLDEN SECTION: A MAP FOR EARLY ARTISTS

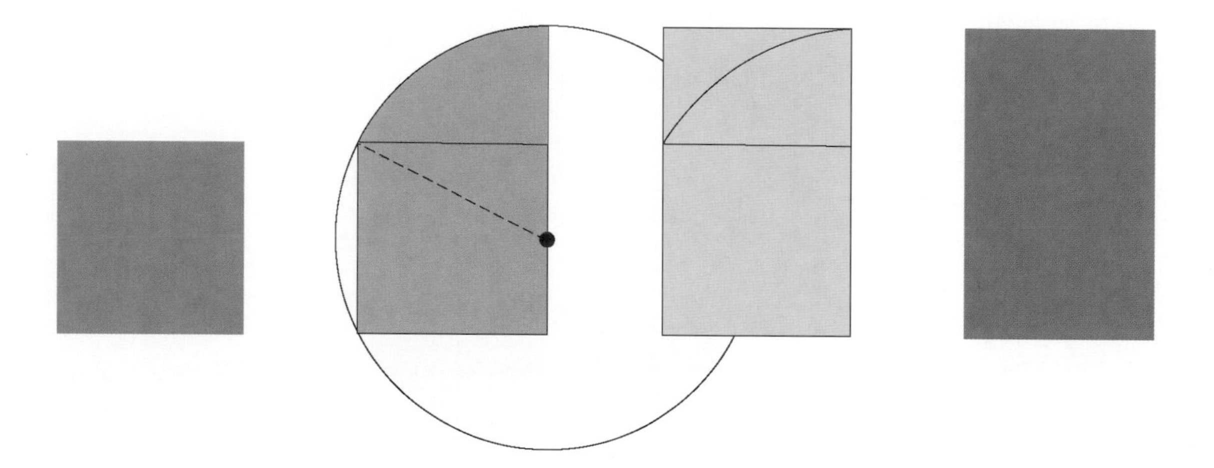

CONSTRUCTION OF A GOLDEN RECTANGLE

Step One

Construct a simple square.

Step Two

Using a line from the midpoint of one side of the square to an opposite corner as the radius, construct an arc that will define the extents of the new rectangle.

Step Three

Using the endpoints of the arc, complete the rectangle.

Step Four

This is the proportion recognized since antiquity as the Golden Section or Golden Rectangle.

THE GOLDEN SECTION IS USED TO DIVIDE A CANVAS ACCORDING TO OVERALL DESIGN AND CONTENT

There are many considerations for an artist in the creation of a painting. In order to balance the elements of color, movement, and content within the shape of the canvas, key attention points or shapes are placed in certain relationships with other key points. Whether this is done according to some inner sense of harmony and balance or it is calculated is something only the artist will know. As observers we can appreciate the final result by overlaying shapes of Divine Proportion to increase our awareness of the overall dynamics.

Geometer by a medieval artist

Salaman and Absal on the Heavenly Isle painted for the Persian Sultan Ibrahim Mirza

Self-portrait by Rembrandt (1606–1669)

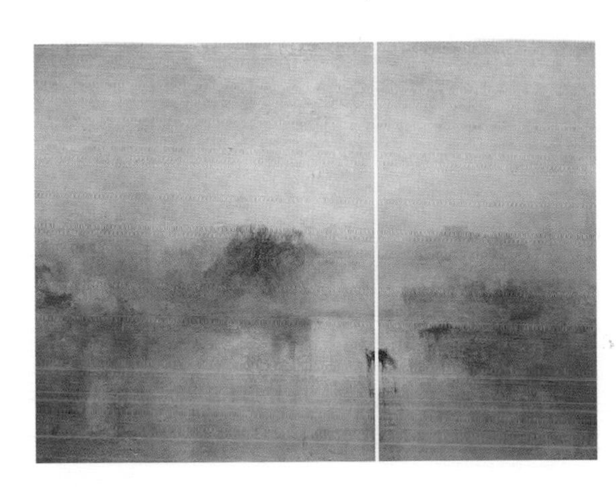

Norham Castle at Sunrise by J.M.W. Turner (1775–1851)

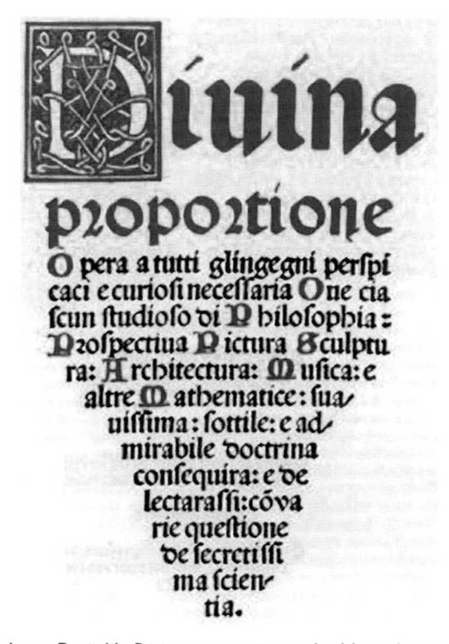

Diuina
proportione
Opera a tutti glingegni perſpi
caci e curioſi neceſſaria Oue cia
ſcun ſtudioſo di Philoſophia:
Proſpectiua Pictura Sculptu
ra: Architectura: Muſica: e
altre Mathematice: ſua-
uiſſima: ſottile: e ad-
mirabile doctrina
conſequira: e de
lectaraſſi: cōva
rie queſtione
de ſecretiſſi
ma ſcien-
tia.

Luca Pacioli's *Divina proportione* relied heavily on the work of Piero della Francesca. The book is also notable for its pictures of the regular solids, drawn by Leonardo da Vinci during the period in which he studied with Pacioli.

This 1994 Italian stamp portrays Luca Pacioli in a fashion that resembles his pose in the painting by Jacopo de Barbieri.

connections and Pacioli studied theology and became a friar in the Franciscan Order.

He spent his time traveling and teaching and wrote two more books before beginning the first of his two most famous, *Summa de arithmetica, geometria, proportioni et proportionalita*, a summary of all the mathematics known at that time.

Ludovico Sforza, regent to the duke of Milan at that time, desired to make his court the finest in Europe and invited Leonardo da Vinci to work as a court painter and engineer. When Ludovico became duke, Pacioli was invited to teach mathematics at that court as well, a move that may have been made at the prompting of Leonardo, who was enthusiastic to learn all he could of mathematics.

Pacioli and Leonardo became close friends, discussing both art and mathematics at length, both gaining greatly from the other. It was at this time that Pacioli began work on the second of his two famous works, *Divina proportione*, with figures drawn by Leonardo. *Divina proportione* comprised three books and was a study of the Divine Proportion. In it Pacioli looked at the theorems of Euclid that relate to this ratio as well as regular and semiregular polygons. The second book discussed the importance of the Divine Proportion in architectural design, and the third was a translation into Italian of one of Piero della Francesca's works.

Meanwhile, Louis XII became king of France and, being a descendant of the first duke of Milan, he claimed the duchy. The French armies entered Milan and captured Ludovico Sforza when he attempted to retake the city a year later. Pacioli and

Leonardo fled together to Mantua, where they were the guests of Isabella d'Este, and then went to Venice. From Venice they returned to Florence, where Pacioli and Leonardo shared a house for a while and Pacioli spent his time teaching and discussing mathematics until he died in 1514 at about age 70.

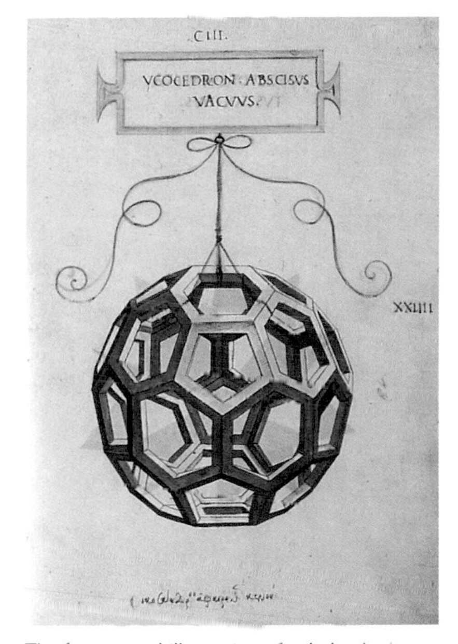

The first printed illustration of a dodecahedron was drawn by Leonardo da Vinci.

Giorgio Vasari's 1550 biography of Piero della Francesca accused Pacioli of plagiarism and claimed that he stole Piero's work on perspective, arithmetic, and geometry. This was an unfair accusation, because although Pacioli relied heavily on the work of others, creating little that was original, he never attempted to claim the work as his own and acknowledged whatever sources he used.

In his book *Divina proportione* Pacioli attributes five qualities of God to the Divine Proportion. The first four qualities are unity and uniqueness, trinity, the impossibility of being defined in human terms, and immutability. Of the fifth attribute, he says:

> *As God breathes life into the cosmos through the "fifth essence" and to the four earthly elements and to every thing in nature, so our Divine Proportion breathes life into the Dodecahedron.*

The dodecahedron is one of Plato's five solids. It is made of twelve pentagons and to Plato it represented the cosmos. An illustration from the book, drawn by Leonardo da Vinci, appears above.

There were three great artists who were of enormous influence in Pacioli's life: Leone Battista Alberti (1404–1472), Piero della Francesca (ca.1422–1492), and Leonardo da Vinci (1452–1519).

Pacioli's Inspirations

The Flagellation by Piero della Francesca (1422–1492)

Piero della Francesca is now recognized as one of the great Renaissance painters, although his painting had little influence on his contemporaries. He came from a family of fairly prosperous merchants and in his own time he was known as a highly competent mathematician. He enjoyed great fame for his writings on perspective, and in 1497, not long after Piero's death, Pacioli described him as "the monarch of our times of painting and architecture." Two generations later he was highly praised by Giorgio Vasari.

Of the books he wrote, three are now known to survive. The titles by which they are known are: *Abacus Treatise* (*Trattato d'abaco*), *Short Book on the Five Regular Solids* (*Libellus de quinque corporibus regularibus*), and *On Perspective for Painting* (*De prospectiva pingendi*).

On Perspective for Painting is one of the very earliest works of his time to deal with the mathematics of perspective. Piero was determined to demonstrate that his techniques were firmly based on the science of vision, for the work was conceived as a manual for teaching painters to draw in perspective. There are many diagrams and illustrations, but unfortunately none of the known manuscripts has illustrations actually drawn by Piero himself.

None of Piero's mathematical work was published under his own name during the Renaissance, but it seems to have circulated quite widely in manuscript form and became influential through its incorporation into the works of others. Much of his work appears in Pacioli's *Divina proportione*.

Leone Battista Alberti (1404–1472), was an architect, humanist, antiquarian, mathematician, art theorist—the "universal man" of the early Renaissance. His influential treatise *Della pittura* (*About Painting*) was the first modern manual for painters. It was circulated in manuscript until 1540, when it was first printed.

> *Let no one doubt, that the man who does not perfectly understand what he is attempting to do when painting, will never be a good painter. It is useless to draw the bow, unless you have a target to aim the arrow at.*
>
> *Nothing pleases me so much as mathematical investigations and demonstrations, especially when*

I can turn them to some useful practice drawing from mathematics the principles of painting perspective and some amazing propositions on the moving of weights.

In 1452, the year when the German Johann Gutenberg invented the printing press, Alberti similarly discovered a way of tracing natural perspectives, as well as a method of reproducing small objects on a large scale.

Melancolia by Albrecht Dürer (1471–1528)

In the autumn of 1506 Dürer rode from Venice to Bologna, to the home of Luca Pacioli in order to be initiated in the mysteries of a "secret perspective."

In Dürer's woodcut *Melancolia* there is a magic square in the background, geometric solids, and the sun's rays, all acting as lines of projection.

Device invented by Dürer to draw proportion

Dürer experimented with optics and studied nature assiduously. He is generally regarded as the greatest German artist of the Renaissance.

...geometry is the right foundation of all painting.
ALBRECHT DÜRER

Engraved portrait of Leonardo da Vinci by Giorgio Vasari (1511–1574)

Mona Lisa

Artists such as Alberti, Piero della Francesca, and Leonardo da Vinci all studied engineering, mathematics, science, and architecture while pursuing their art. They all wrestled with the problem of how to paint realistic three-dimensional scenes on two-dimensional canvases and are all respected for the work they did in developing techniques to achieve this end. They reasoned that if they perceived a scene outside through the flat plane of a window, they should be able to translate the scene onto their canvases. A variety of techniques were developed that allowed them to do this.

Leonardo da Vinci

Leonardo da Vinci was the archetype of the Renaissance man and perhaps more than any other figure epitomizes the Renaissance humanist ideal. His *The Last Supper* and *Mona Lisa* are among the most popular paintings of the Renaissance, and his notebooks reveal a spirit of scientific inquiry and a mechanical inventiveness that were centuries ahead of their time.

Leonardo's parents were unmarried at the time of his birth. His father was a notary and landlord, and his mother was a young peasant woman who married someone else shortly thereafter. Leonardo grew up on the estate of his father's family where he was treated as a legitimate son and received the usual education of that day.

His artistic inclinations must have appeared early, for his father apprenticed him to the artist Andrea del Verrocchio around age fifteen. At twenty Leonardo was accepted into the painters' guild of Florence, but he remained in his teacher's workshop for five more years and then worked independently in Florence for another ten years. At that point Leonardo moved to Milan to work in the service of the duke—a surprising move, for he had just received his first substantial commissions from his native city of Florence for two works that were never begun. He must have had other reasons for leaving Florence; it is likely that he was enticed by Duke Ludovico Sforza's brilliant court and the meaningful projects awaiting him there.

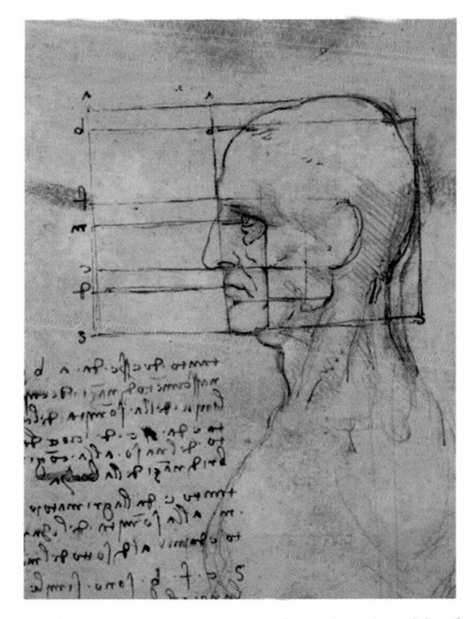

Study of the proportions of the head and body

Cross section of branches

Leonardo spent seventeen years in Milan, until Ludovico's fall from power in 1499, and he was constantly kept busy as a painter, sculptor, and designer of court festivals. He also acted as technical adviser in architecture, fortifications, and military matters and served as a hydraulic and mechanical engineer.

As a painter, Leonardo completed six works in those years in Milan including *The Virgin of the Rocks* and *The Last Supper*. During this period Leonardo also worked on a grandiose sculptural project that seems to have been the real reason he was invited to Milan—a monumental equestrian statue in bronze to be erected in honor of Francesco Sforza, the founder of the Sforza dynasty. Leonardo devoted twelve years to this task and in 1493 the sixteen-foot-high clay model of the horse was put on public display and preparations were made to cast it. However, war was imminent, and the metal ready to be poured for the statue was used to make cannons instead. The project came to a halt and he never completed it.

When he and Pacioli arrived in Florence Leonardo was received with acclaim. He spent several years as a military architect and general engineer traveling across the country as a surveyor and then returned to Florence for a period of intense scientific study. He did dissections in the hospital to broaden his understanding of the structure and function of the human body. He made

Bes, a dwarf-like Egyptian God who helps mothers in childbirth, plays the harp.

systematic observations of the flight of birds and engaged in studies of the movement and properties of water and air. His investigations became increasingly driven by a conviction that force and motion operate in accordance with orderly, harmonious laws.

At the age of sixty-five Leonardo moved to France where he spent the last three years of his life in a small house on the Loire and was given the title *Premier peintre, architecte et méchanicien du Roi* ("First painter, architect, and engineer to the King"). His last years were spent editing his scientific studies, a treatise on painting, and another on anatomy.

Music, Harmony, and Proportion

The study of mathematics, geometry, and the principles of proportion was naturally interesting to artists of the Renaissance. Although they were not musicians, they could not but ponder the exquisite way in which proportion figured into sound, and more significantly, into music.

In his *Ten Books of Architecture*, Alberti writes that proportion within a geometrical figure, a musical scale, or, indeed, a mathematical sequence, can be said to be "a harmonious relationship between the parts, with and within the whole."

He goes on to propose that beauty is an agreement between the different parts of a whole, an agreement in which the parts and the whole are all in accord with the "principal Law of Nature."

> *The Ancients...did in their Works propose to themselves chiefly the Imitation of Nature, as the greatest Artist at all Manner of Compositions.*

To Alberti, the Law of Nature was a very exact term and he proceeded to develop an explanation for the way Nature's Law operates in our world of shapes and forms and to apply the Rule of Proportions. Then he named the Musical Masters as those who understood this rule the best.

A bard or ballad singer from ancient Greek culture playing a lyre

> *The Rule of these Proportions is best gathered from those Things in which we find Nature herself to be most complete and admirable; and indeed I am every day more and more convinced of the Truth of Pythagoras's Saying, that Nature is sure to act consistently, and with a constant Analogy in all her Operations:*
>
> *From whence I conclude that the same Numbers, by means of which the Agreement of Sounds affects our Ears with Delight, are the very same which please our Eyes and Mind. We shall therefore borrow all our Rules for the Finishing our Proportions, from the Musicians, who are the greatest Masters of this Sort of Numbers, and from those Things wherein Nature shows herself most excellent and complete.*

This linking of nature's perfect sense of proportion in both art and music is consistent with a long tradition of philosophical thought regarding music dating back to Classical Greece.

From historical accounts, it is clear that music has always held the power to move people; its ecstatic potentialities have been recognized in all cultures and have at times been both exploited and strictly restricted.

Moroccan illustration from the 13th century of the story of Bayad and Riyad: "Bayad singing and playing the 'ud before the Lady and her Handmaidens"

Playing the Flute by Pine and Stream by Qiu Ying (1495–1552)

In India, music served religious ceremony from the earliest times; Vedic hymns are among the earliest known religious expressions. Though the art of music developed over many centuries into a profound melodic and rhythmic intricacy, a religious text or the guideline of a story always determined the structure. Chinese music has also traditionally been joined to ceremony. Confucius (551–479 B.C.E.) assigned an important place to music in the service of a well-ordered universe. He believed that no man is equipped to govern unless he can understand music. Music, Confucius thought, reveals character through the six emotions that it can portray: sorrow, satisfaction, joy, anger, piety, and love. According to Confucius, great music is in harmony with the universe, restoring order to the physical world through its harmony. Music, as a true mirror of character, makes pretense or deception impossible.

Looking at our own cultural roots we know that music was an important aspect in the life of ancient Greeks, but we have no idea how that music actually sounded. Only a few notated fragments have survived, and no key exists for restoring them.

The Greeks were as given to theoretical speculations about music as they were to all other aspects of life. They had a system of notation for it, and so were equipped to "practice music." The Greek term from which the word music is derived was a global one, referring to any art or science practiced under the aegis of the Muses. Music, therefore, was all-encompassing. Pythagoras, having laid the foundations for the scale we use, discovered the relationship between the pitch of a note and the length of a string.

Plato, like Confucius, believed there was an ethical component of music and was as anxious to see it regulated because of its supposed effects on people. Plato drew a direct correspondence between a person's character and the music he listened to. In the *Laws*, he declared that rhythmic and melodic complexities were to be avoided because they led to depression and disorder.

He also recognized that music is an echo of the divine harmony, and believed that because rhythm and melody imitate the movements of heavenly bodies, music must be related to the movement of the spheres and the intrinsic order of the universe. However, Plato so distrusted the emotional power of music that he thought it was necessary to impose a strong censorship on its performance. He concluded that the sensuous qualities of music are actually dangerous, and while he admitted and valued music in its idealized and divine form, he had grave concerns about music's "earthly" and actual effects.

Confucius (551–479 B.C.E.) is ancient China's highly regarded teacher and political analyst. His philosophy of unity and stability can be seen as a reaction to the turbulent times in which he lived.

Plato's influence on music was to be dominant for at least a millennium. The conservative aspects of his philosophy, with its inherent fear, were conducive to the maintenance of order, and the restricted role of music is nowhere more clearly illustrated than in the history of Christianity. In the Church's plainchant, melody was used for textual illumination only, and the configurations of sound took their cue from the words. St. Augustine (354–430), who loved music and valued its usefulness to religion, was also fearful of the sensuous element and anxious that the melody should never take precedence over the words.

Medieval cathedrals were designed in such a way that because of their enormous height there is a delayed action of the voice going out and coming back—a construction that enhances chanting. When plainchant is sung it is as if angels are responding. The sounds literally come to life, and music and human bodies

An angel musician by Fra Angelico (ca. 1400–1455)

become one great, staggering phenomenon, rising up in rejoicing song and falling back down like golden rain.

The German astronomer Johannes Kepler (1571–1630) perpetuated Pythagoras' idea of the harmony of the spheres, and attempted to relate the divine aspects of music to planetary movement. René Descartes (1596–1650) also realized the perfect mathematical expression of music, but expressed concerns about its imaginative, exciting, and hence immoral, effects. Immanuel Kant (1724–1804) ranked music as lowest in his hierarchy of the arts. What he distrusted most about music was its wordlessness; he considered it useful for enjoyment but negligible in the service of culture. Allied with poetry it may, he thought, acquire conceptual value.

Before the 19th century, musicians, unlike painters and architects, were seldom theorists; geniuses of music, such as Johann Sebastian Bach, produced not learned treatises but great monuments of their art.

Our present-day experience of music owes much to two German philosophers, Arthur Schopenhauer (1788–1860) and Friedrich Nietzsche (1844–1900), who brought a new concept to the theory of music, though articulated in different ways. Both saw in music an art that is not "objective" in the way that other arts are. It is alive. It does not sit on a canvas and has no physical structure.

While the creation of art is always immediate, the experience of music for the listener is directly connected to the creative process of the musician. Until recent times in which recordings could be made, music—like dance, drama, and the spoken word (all of which are often integrated with music)—was performed, thus there is a more direct connection of the listener to the emotion of the musician.

Humankind has always acknowledged the connection between art and human feeling, but what music shares more immediately than other arts is the fact that it cannot be seen. It exists, like emotions, in some nonphysical realm. Music is everywhere to be heard, but it can't be touched. What is music?

Sounds, whether in the form of noise or music, are the result of vibrations. Once an object begins to vibrate, the vibrations cause surrounding molecules of air to also vibrate.

Every object that makes noise has a method of creating sound in a different way. For instance, when a violin string is plucked its movement causes the other strings and the entire instrument to vibrate. These vibrations travel outward from the source in three dimensions and reach our eardrums. From deep within the ear, a signal is sent to our brain to give us the sensations of sound.

Back in the days of ancient Greece, Pythagoras observed that when a blacksmith struck his anvil, different notes were produced according to the weight of the hammer. He then went on to discover that mathematical ratios relate different vibrations and thus he was the first to associate music and mathematics. His experiments proved that the ratio of the number of vibrations of a plucked string to a string half its length is exactly one to two.

In today's language, we understand that if a plucked string produces a certain note with a particular number of vibrations (say 264) then when the string is depressed at half its length it vibrates at exactly twice the speed (say 528).

Pythagoras testing the relationships of music and numbers in a woodcut from a 1492 book by Franchinus Gaffurius called *Theorica musicae*

First-century relief from Greece of a muse playing on a lyre

As sweet and musical
As bright, Apollo's lute, strung
with his hair;
And when Love speaks, the voice
of all the gods
Makes heaven drowsy with the harmony.
SHAKESPEARE

The difference between these two notes establishes the range of a scale. The scale we use we call an octave (from the Latin for eight) and it has seven distinct notes to which we have assigned the seven letters from A to G. The eighth note is an octave away. Every sound, or note, has a particular frequency or number of vibrations per second and there is a one-to-one correspondence between sounds and number.

The number of notes or subdivisions in a scale is arbitrary and is influenced by a number of factors including the design of instruments and the way they are played. Although our ears are not designed to discern all the possible distinct sounds, a trained ear can hear about 300 different notes or sounds or intervals in one octave.

After experimenting with plucked strings the Pythagoreans discovered that the intervals that most pleased people's ears were

octave	1 : 2
fifth	2 : 3
fourth	3 : 4

The Pythagoreans taught that the orbit of each of the seven planets produces a particular note according to its distance from the still center that they thought was Earth. This idea came to be known as *Musica Mundana*, Music of the Spheres or Harmony of the Universe, and the sounds produced were thought to be so exquisite and rarified that our ordinary ears are unable to hear them. According to Philo of Alexandria, Moses heard this Cosmic Music when he received the Tablets on Mount Sinai; St. Augustine believed humans hear it on the point of death.

The Pythagorians also believed that different musical modes have different effects on the person who hears them, and it is said that Pythagoras once cured a youth of his drunkenness by prescribing a melody in a certain mode and rhythm. At certain healing ceremonies in Greece patients underwent therapies accompanied by music. The Roman statesman, philosopher, and mathematician, Boethius, has explained that the soul and the body are subject to the same laws of proportion that govern music and the cosmos. We are happiest when we conform to these laws, he said, because of the harmonious resonance this sets up inside us.

Harp player from Ancient Egypt singing to Horus, God of the Sun

Plato described a reality in which Being and Existence are bound together as One; in which reality is inseparable from the natural harmony and proportion that exist throughout creation; and the relationship of all parts, within the whole and to the whole, is reflected in the relationship between us and our universe.

Thus, to be creative, be it architectural, artistic, musical, or even agricultural, came to be seen as a natural consequence of a person's awareness of the harmonious relationship of all aspects of existence.

> *The painter's mind is a copy of the divine mind, since it operates freely in creating the many kinds of animals, plants, fruits, landscapes, countrysides, ruins, and awe-inspiring places.*
>
> LEONARDO DA VINCI

Chapter 5

Divine Proportion in Nature

Living in the world without insight into the hidden laws of nature is like not knowing the language of the country in which one was born.

HAZRAT INAYAT KHAN

THE ARTIST WITHIN EACH OF US draws inspiration from any number of places—without and within—and usually we acknowledge nature to be one of our greatest teachers. The immutable laws that play out in the unfolding of life are as remarkable in their simplicity as they are in their complex subtleties, and we are intimately connected with them. In the wind and sand, in the flow of water and the changing colors of dawn and dusk, nature's laws are there for all to see.

"Heaven is under our feet as well as over our heads," said Henry David Thoreau; and "Nature is full of genius, full of the divinity, so that not a snowflake escapes its fashioning hand."

Drawing of an apple by William Hooker (1779–1832)

The apple, one of the most popular and widespread fruits that actually belongs to the rose family, has played an important part in mythology and folklore. Apples were often thought to be the food of gods and represent life itself. In the famous legend of Adam and Eve, apples grew on the tree of forbidden fruit and represent sacred knowledge.

OPPOSITE: *Insects and Flowers*, an anonymous Chinese painting

FRACTALS: THE DIVINE GEOMETRY OF NATURE

EUCLIDIAN GEOMETRY was concerned with an abstract perfection almost nonexistent in nature. It could not describe the shape of a cloud, a mountain, a coastline, or a tree. As Mandelbrot said in his book *The Fractal Geometry of Nature*:

Clouds are not spheres, mountains are not cones, coastlines are not circles, and bark is not smooth, nor does lightning travel in a straight line.

When the data of fractals are colored so that certain values get certain hues, the computer-generated solutions form shapes with an eerie beauty.

Big whorls have little whorls,
Which feed on their velocity;
And little whorls have lesser whorls,
And so on to viscosity.

LEWIS RICHARDSON (1881–1953), A MATHEMATICIAN WHO STUDIED WEATHER PREDICTIONS

So, Nat'ralists observe, a Flea
Hath smaller Fleas that on him prey,
And these have smaller fleas to bite 'em,
And so proceed ad infinitum.

JONATHAN SWIFT (1667–1745)

The principle of self-similarity is shown here with each sphere being half the size of the one it attaches to. As smaller spheres are added the total surface area approaches infinity while the volume remains finite.

FRACTALS HAVE COME TO BE REFERRED TO as the geometry of nature. They are shapes or behaviors, often associated with beautiful computer graphics, that have similar properties at all levels of magnification. The term fractal is used to describe a particular group of irregular shapes that do not conform to Euclidean geometry, and their most well-known attribute is self-similarity, meaning that they appear to have copies of themselves buried deep within the original. They also have infinite detail. Just as the sphere is a concept that includes raindrops, basketballs, and Earth, so fractals are a concept that unites clouds, coastlines, lightning bolts, and trees.

Discovered and named by Benoit Mandelbrot, now both an IBM scientist and professor of mathematics at Yale, fractals evolved out of answers to two questions that arose before the invention of computers. The first of these questions was posed when mapmakers were measuring the length of Britain's coast. Looking at a 1:100 map they obtained one measurement. When subsequently analyzing a 1:1000 map they found that the length of the coast had significantly changed and the resulting measurement was longer. They realized that the closer they looked, the more detailed and longer the coastline got.

The second question arose when the French mathematician Gaston Julia (1893–1978) wondered what complex algebraic formulas would look like when drawn. Specifically, he was looking at formulas that used an imaginary number (i)—described mathematically as the square root of minus one (√-1).

Taking these two questions together, we begin to see what is often called the geometry of the dimensions. Most of us are familiar with the fact that a point is zero-dimensional, a line is one-dimensional, a plane is two-dimensional, and a cube is three-dimensional. Fractals, describing things like a coastline or a lightning bolt, lie somewhere between these integral numbered dimensions. They are fractional dimensions—hence their name, fractals. The word fractal also had a root in the Latin adjective *fractus* and the corresponding Latin verb *frangere*, which means "to break" or "to create irregular fragments."

Benoit Mandelbrot was the first person to get computers to do the many repetitive calculations that made the well-known images of fractals appear. His early work in the 1950s and 1960s suggested that the variations in stock market prices, the probabilities of words in English, and the fluctuations in turbulent fluids all had something in common.

As a student, Mandelbrot never properly learned the alphabet and barely learned the multiplication tables, but he had a special genius—an extraordinarily visual mind—and solved problems with great leaps of intuition rather than with the established techniques of logical analysis. After obtaining a doctoral degree in math he turned to research.

His famous study in economics concerned the price of cotton, a commodity that has a supply of data going back hundreds of years. While he was aware that the day-to-day price fluctuations were unpredictable, his computer analysis provided an overall pattern that was revolutionary. He saw that the daily unpredictable fluctuations repeated themselves over larger, longer scales of time. He discovered a relationship of symmetry between the long term price fluctuations and the short-term ones.

Fractals

The word fractal was coined in 1975 by IBM mathematician Benoit Mandelbrot to describe an intricate looking set of curves. Before the advent of computers, with their ability to quickly perform massive calculations, many of these curves had never been seen. Fractals often exhibit the principle of self-similarity, which means that various copies of an object can be found in the original object at smaller size scales.

The Golden Spiral— One of Nature's Special Blueprints

The spiral, an essential tool in nature's palate, has long been regarded as one of the most significant. Look into the sky and spirals are there, look at the water, look into the wind. Peel off the leaves of a head of lettuce or a cabbage, peer closely at the seeds in a daisy. Spirals are everywhere—some obvious, some less so. From embryos to galaxies, the spiral offers what is perhaps one of nature's most dynamically proportionate messages in that it arises out of the reconciliation of opposites, harmonious and unaligned. It is the middle way, the path of least resistance, never leaning too far in one direction or the other but always finding perfect balance.

Many kinds of spirals have been studied and described in the language of mathematics, the first dating from the days of ancient Greece. All spirals have in common the fact that they unwrap around a fixed point at a changing distance. In other words they unwind around a point while moving ever farther from the point.

The Golden Spiral, the one that is based upon the Divine Proportion, is what mathematicians call a logarithmic spiral. Its growth is as perfect as all other aspects of Divine Proportion, for embedded in this spiral are all the beautiful mysteries of Φ's harmony and balance. Intrinsic to its pattern of growth or decay is the perfect ratio that says the whole is to the larger as the larger is to the smaller.

SPIRALS

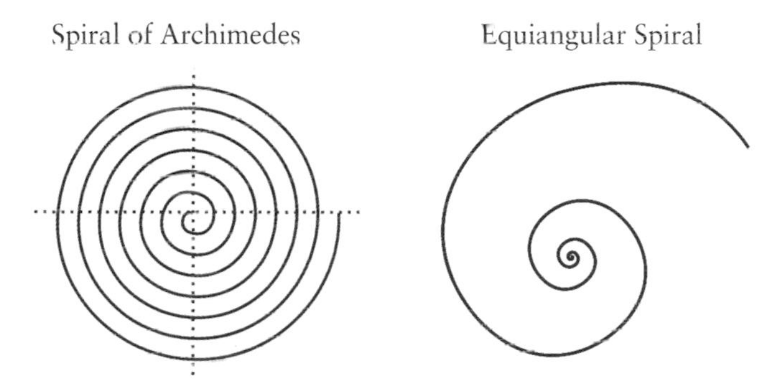

THE TWO SPIRAL SHAPES THAT SHOW UP MOST FREQUENTLY IN NATURE are the Archimedean and logarithmic spirals.

Archimedean Spiral

This spiral was studied by Archimedes (ca. 287–212 B.C.E.) in his work *On Spirals*. Archimedes was a Greek mathematician who described the spiral that bears his name as one whose distance from the point grows at a fixed rate. In mathematical terms it is a curve in which polar coordinates (r, q) can be described by the equation r = a + bq, where a and b are real numbers. Changing parameter a will turn the spiral, while b controls the distance between the arms. This is the spiral we see in a coil of rope, a clock spring, record grooves, and a roll of paper. The "helix" is a three-dimensional version of this spiral, and it is found in coil-springs, bolts, barber poles, and the double helix of our genetic heritage in the DNA molecule.

Logarithmic or Equiangular Spiral

The equiangular spiral was so named by Descartes, and its properties of self-reproduction were described by Jakob Bernoulli. The logarithmic spiral describes a family of spirals whose curve cuts all radii vectors at a constant angle. Like the spiral of Archimedes it grows from a fixed point. Its equation, which is a little more complicated, says

$$r = ae^{b\theta},$$

where r is the distance from the origin, q is the angle from the x-axis, a and b are arbitrary constants, and e is a mathematical constant that represents the number 2.71828....

This spiral is also known as the growth spiral or *spira mirabilis*.

The logarithmic spiral can be distinguished from the Archimedean spiral by the fact that the distances between the arms of a logarithmic spiral increase in geometric progression, while in an Archimedean spiral these distances are constant.

The Swiss mathematician Jakob Bernoulli (1654–1705), one of the founders of calculus, devoted a great deal of time to questions about spirals and wrote a treatise called *Spira mirabilis* (*Wonderful Spiral*).

Having spent many hours on the study of spirals he found the properties of the logarithmic spiral to be almost magical, and he requested that one be carved on his tombstone with the Latin inscription *Eadem Mutata Resurgo,* meaning "I shall arise the same though changed."

Unfortunately, the spiral that was finally inscribed there was the Archimedean spiral and not the logarithmic one that he so loved.

Chambered Nautilus

As we saw in Chapter 1, the Golden Spiral can be geometrically constructed in two different ways—using Golden Triangles and Golden Rectangles.

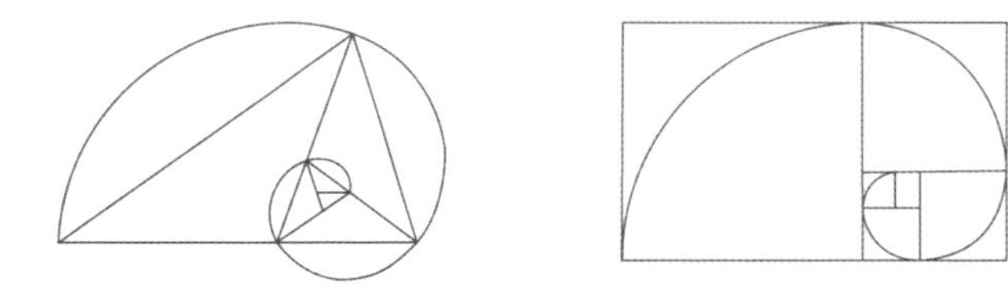

The Golden Spiral can also be created using the Fibonacci numbers: 1,1,2,3,5,8,13,21.... Beginning with two small squares of one unit each, place next to them a square of two units, and next to this one of three units, and then one of five... continuing to grow the spiral.

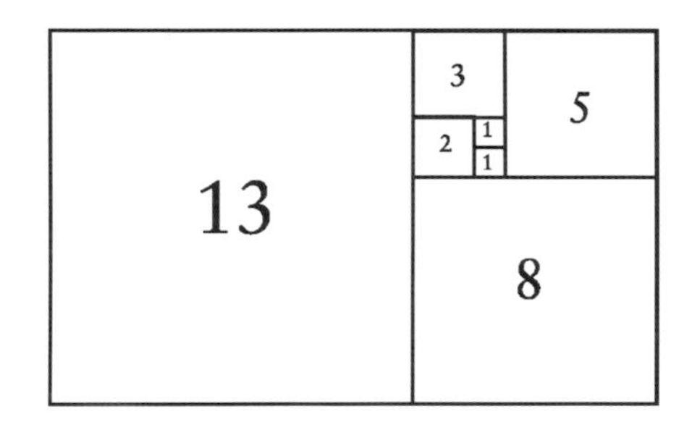

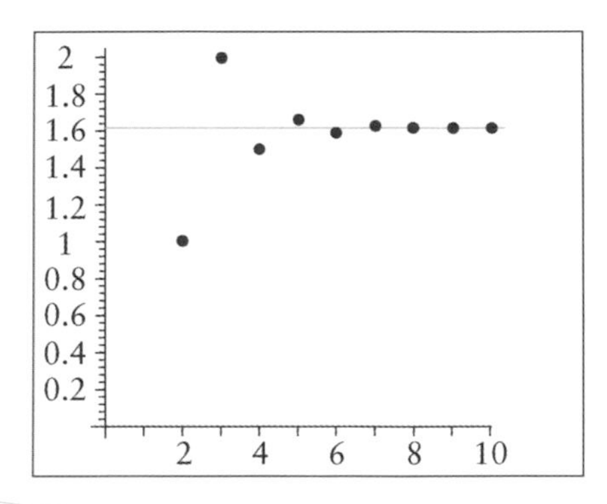

This graph shows how the sections of the Fibonacci Spiral approach Φ, the Divine Proportion, or 1.618....

1/1 = 1.000000
2/1 = 2.000000
3/2 = 1.500000
5/3 = 1.666667
8/5 = 1.600000
13/8 = 1.625000
21/13 = 1.615385
34/21 = 1.619048

55/34 = 1.617647
89/55 = 1.618182
144/89 = 1.617978
233/144 = 1.618056
377/233 = 1.618026
610/377 = 1.618037
987/610 = 1.618033

Creating a spiral using squares based on Fibonacci numbers does not produce a true mathematical spiral. The Fibonacci spiral is made up of fragments (parts of circles) and does not go on getting smaller and smaller because it stops at the unit one. However, it is a good approximation of the Golden Spiral, and the principle of Fibonacci numbers increases our understanding of the growth pattern of the Golden Spiral by adding what we know about the Fibonacci principle: Any number in the series is found by adding the two previous numbers.

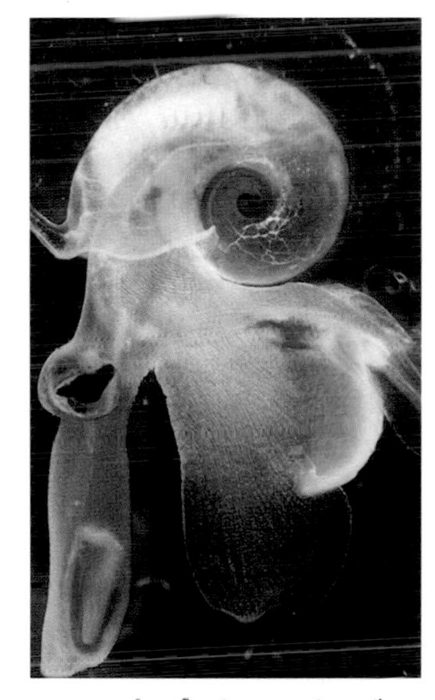

An oxygyrus, a free-floating oceanic snail

Spirals in Nature

Drawing a Golden Spiral or a Fibonacci Spiral on paper only begins to describe the spiral that nature creates—one that embodies the dynamic principles of regeneration and whose imprint in life is of symmetrical and balanced growth. Life is either expanding, growing, being drawn out; or it is diminishing, dissolving, collapsing. Seeing the beauty of an unfolding leaf or the patterning of petals in a rose we immediately recognize the perfect and delicate spiral. We see it in the water and in the clouds, wonderful patterns that dissolve and reappear out of nothing. And nature's attention to detail is so wondrous we see the same spiral shapes repeated in our own bodies, unfurling our own lives in the same unfolding patterns we see in plants.

The Eye of the Storm

Mathematicians call the Golden Spiral's eye an *asymptote*, a place always approached but never reached. The calm eye of a storm is the center of gravity around which wind and water are expanding and contracting, around which the whole storm is balanced. The eye of every spiral is a dynamic place where all opposites meet and where life and death are one phenomenon. All the forces that create growth and keep it in balance are at work in the eye, the source.

Hurricane Bonnie approached the eastern seaboard of the United States on August 25, 1998—a towering mass of spiraling cloud twice as tall as Mount Everest, unusually high for an Atlantic storm.

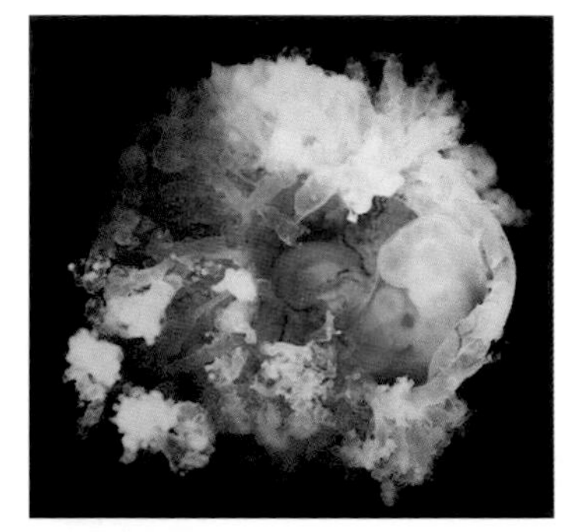

Human embryo

Spirals in weather and ocean patterns

In its movement through space it has been proven that Earth's rotation is slowing down very slightly over time, about one second every ten years. This rotation is minutely altered by the effects of spiraling weather and ocean patterns. During the period 1982–1983, Earth slowed by 2/10,000 of a second due to that year's El Niño, which caused large enough changes in the ocean water temperature and air pressure that worldwide weather conditions were significantly affected. Dr. Dennis McCarthy of the United States Naval Observatory reported on January 24, 1990, that Earth's day was lengthened by 5/10,000 of a second.

The eye of the storm is in fact a safe place, protected by a wall of clouds. Hurricane survivors talk of how strange it was to go out and view the eye of the storm—blue sky overhead surrounded by a wall of angry clouds. Ships at sea have been known to survive hurricanes by riding them out in the eye of the storm.

Balance-in-Motion

The great whirlpool of movement that we see in our solar system incorporates another of the principles that underlies spirals called balance-in-motion. Watching a stick thrown into a whirlpool, we see a principle in action that keeps it always pointing in the same direction. This same principle keeps the axis of our Earth pointing in one direction while spinning on itself and moving around the sun.

We see this same principle at work when a hawk dives toward its prey in a spiral flight pattern. Since a hawk does not appear to move its eye in its socket, a particular alignment in the retina of the eye allows it to keep moving at a tremendous speed while the image of its prey is locked in place.

All throughout the natural world in everything we observe, from the incredibly vast to the infinitesimally small, we see things moving with different principles of dynamic balance along with an inborn tendency to move toward an equilibrium. Natural systems are self-adjusting and their innate capabilities keep things in balance should a disruptive event occur. The human body also operates on this principle.

Our bodies, too, are filled with spirals. The most obvious of these appear in our ear, but we see them, too, in our fists, in the spiral pattern of cowlicks, in the shape of the human embryo, and even in the structure of our DNA. The spirals in our bodies incorporate many principles, most notably that which Chinese acupuncturists use in their treatments. Working with principles of the whole human map they have learned that all the body parts are reflected on a map drawn on the spiral of the ear.

The cochlea of the human inner ear, the organ responsible for hearing

Inner Balance

Through the pioneering work of Georg von Bekesy (1899–1972) we have begun to understand how we hear. In 1947 he developed a mechanical model of the inner ear. The spiral cochlea (Greek for "snail shell") that sits in the inner part of the ear is the part of the organ that "hears." Its shape corresponds to how musical octaves appear when graphed as wavelengths. Each note is identical to those directly above and below it on the spiral, but with a difference of one octave.

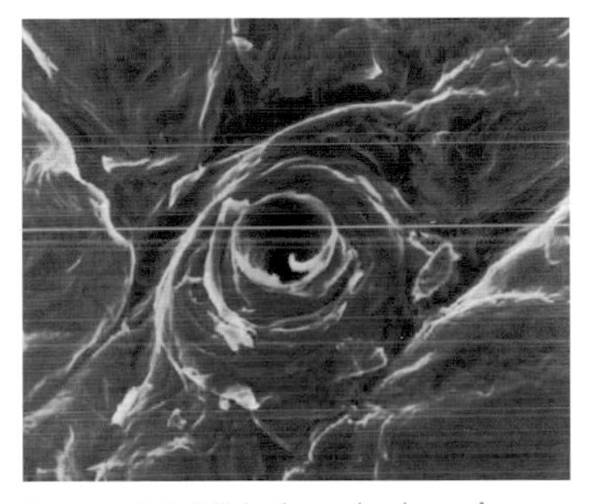

An empty hair follicle shows the shape of a spiral. Except for the palms of the hands and the soles of the feet, all areas of the human skin contain similar follicles from which all our hair grows. Individual hairs last about seven years before being shed and replaced by a new hair growing from the same follicle.

The cochlea inside each ear is encased within a hard shell and filled with liquid. Around the central column is a rail of sensitive hairs that decrease in thickness as the column winds upward. Sound vibrations enter the whirlpool of the outer ear and tap at the cochlea's base, traveling in waves up and around the inner liquid. Each tone makes waves that break at a different point. The turbulence of the breaking wave triggers the sensitive hairs at that location to send electrical signals to the brain for interpretation.

Our human cochlea with its two-and-three-quarter spiral turns allows us to hear approximately ten octaves of sound. Other mammals have different numbers of turns, and this explains why they can perceive different frequencies than humans do.

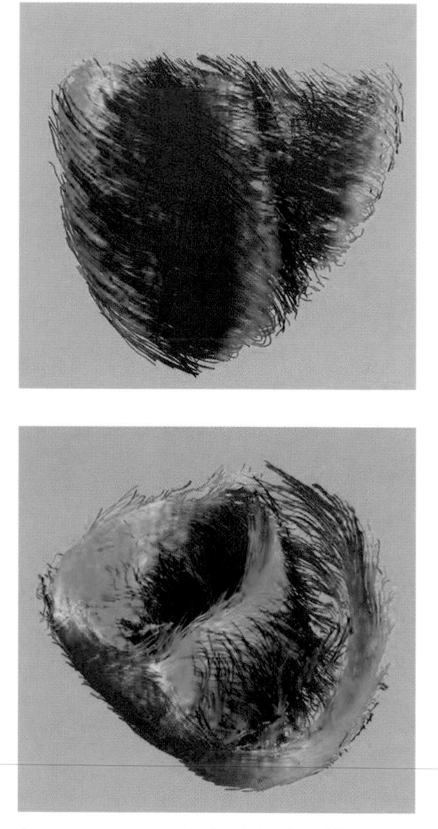

A computer program helped develop this reconstruction of heart muscle fibers in which we see the spiraling shape of the muscle fibers on the inside and outside surfaces of the heart.

Pythagoras' discovery that musical notes can be expressed as mathematical ratios extends to the relationship that is formed when these notes are graphed. The graph forms the shape of the Golden Spiral. When it is stretched into three dimensions it becomes the shape of the cochlea.

Our inner balance depends upon the spiral; it is at work in the shape and functioning of our ears. It is also at work in the endless unfurling spirals that keep every part of our body in a state of regeneration. In the embryo shape that we spiral into in the womb as life begins, the spiral is at work, and at the very center of our body, in our very core, is the human heart—a spiral shaped muscle whose pulsing, through its continuous contracting and relaxing, sends blood rushing through 60,000 miles of blood vessels.

Flow

A vortex is the shape in nature that gathers the energies of wind or water and draws them toward its center. A vortex street is the mathematical description of the flow pattern formed in the wake of an obstacle. One way to create a spiral is to create a vortex street. This can be done by dragging a stick through a pool of water or by pouring one liquid into another. Weather patterns are a result of this phenomenon, as is the process of everything that grows. Whether visible or invisible, vortex patterns surround us. Hot moisture rising off a body of water flows into spirals in the same sort of way that weather patterns flow, leaves unfurl, and shells are formed.

Nature's Spirals of Flow

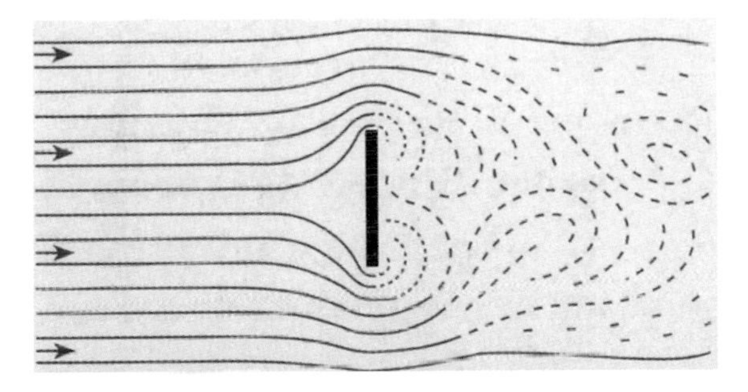

This diagram shows how liquids (including air) form into spirals when met by an obstacle.

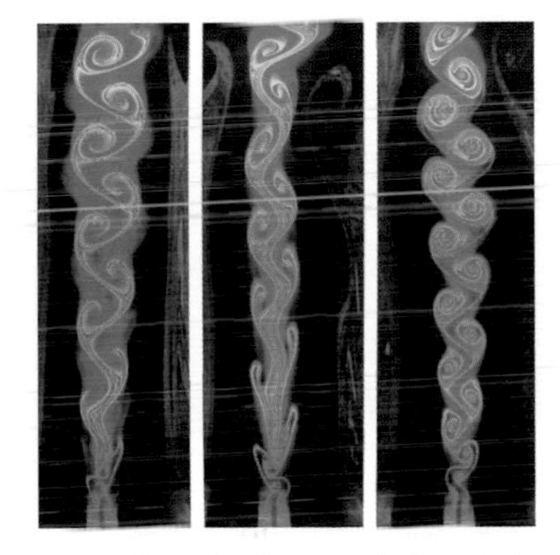

One of the aspects of irregular flow, or turbulence, is that although the patterns are not predictable, there is a certain regularity. These images show the patterns that are formed by changing the speed of two streams of air—one is shown in black and the other in green. Turbulence occurs when the two streams come into contact with each other, and the spiraling pattern is determined by the difference in their speed.

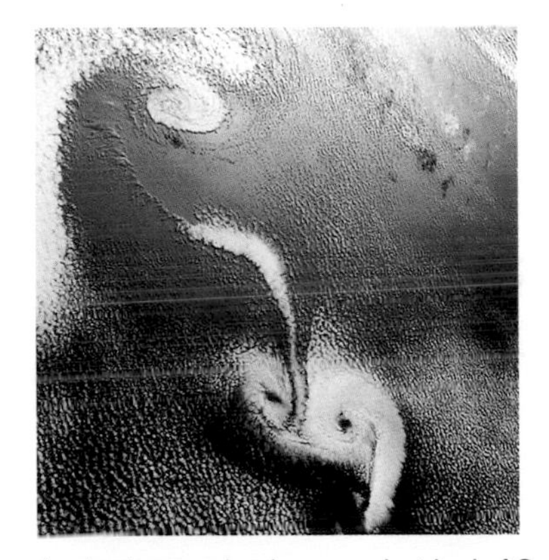

This space shuttle photograph, taken near the island of Guadeloupe in the Caribbean Sea, shows swirling cloud formations in an atmospheric phenomenon known as a Von Karman vortex street. The long line of eddies that rotate alternately clockwise and counterclockwise is caused by obstacles, such as islands, disturbing the airflow as it passes over them. Von Karman vortex streets are often seen in satellite images of clouds.

Two galaxies are colliding 430 million light years away.

Fibonacci's fictitious problem about the breeding of rabbits asks the following: Suppose a newborn pair of rabbits, one male, one female, are put in a field. They are able to mate at the age of one month so that at the end of its second month a female can produce another pair of rabbits. Suppose that the female always produces one new pair (one male, one female) every month from the second month on. How many pairs will there be in one year?

Fibonacci Spirals

Artists, scientists, and philosophers have endlessly studied nature's hand in the unfurling of energies into a spiral pattern. Exploring the mathematical language of spirals reveals another stunning fact about the behavior of the Golden Spiral—a reoccurring adherence to the Fibonacci sequence.

Fibonacci first expressed this sequence as it related to rabbits in the fictitious problem about the breeding of rabbits that we saw in Chapter 3.

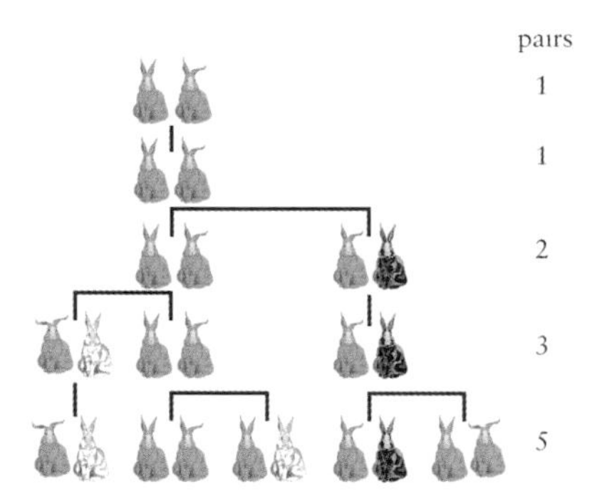

The number of pairs of rabbits that are produced at the start of each month is 1, 1, 2, 3, 5, 8, 13, 21, 34 ... and each new number in the series is formed by adding the two previous numbers.

These numbers—1, 2, 3, 5, 8, 13, 21, 34 ... appear over and over again in certain spiral formations and show up most beautifully in the symmetric arrangement of petals on a rose, where successive leaves grow at an angle of 137.5 degrees—the Golden Angle.

FIBONACCI NUMBERS IN NATURE

The Fibonacci numbers appear remarkably often in nature. When we look at the bud of a rose, we see a striking example. At the center of the rose is the apex, and around the apex the rose's petals emerge, moving away in a circular order. Following their order of appearance, a generative spiral emerges and if we measure the angles between successive petals we find that the angles between them are about 137.5 degrees. This angle is sometimes called the Golden Angle and is found by multiplying 360 degrees by Φ, the ratio formed by successive Fibonacci numbers.

We also see the Fibonacci series in the genealogy of honeybees. Like some other insects, the males (drones) experience a sort of "virgin birth" whereby they are born from unfertilized eggs. The pattern that is established is what maintains the delicate balance of the beehive.

In every colony of honeybees there is one special female called the queen. There are worker bees who are female and are produced by fertilized eggs, and there are drone bees who are male, do no work, and are produced by the queen's unfertilized eggs. These bees have only a mother. They have no father. All the females are produced when the queen has mated and so have two parents.

Female honeybees usually end up as worker bees, but some are fed with a special substance called royal jelly that makes them grow into queens. These bees will eventually start a new colony when the bees form a swarm and leave their home in search of a place to build a new nest.

GENEALOGY OF A MALE DRONE

He has one parent, a female.

He has two grandparents, since his mother had two parents, a male and a female.

He has three great-grandparents: his grandmother had two parents but his grandfather had only one.

Generation	Number of Bees
1	1 drone
2	1 mother
3	2 grandparents
4	3 great-grandparents
5	5 great-great-grandparents
6	8 great-great-great-grandparents

The numbers of bees in successive generations are Fibonacci numbers.

FIBONACCI NUMBERS IN FLOWER PETALS

IN FLOWERS THE FIBONACCI NUMBERS are expressed in a very simple way. Flowers with one and two petals are relatively rare. Flowers with three petals are more common. There are many flowers with five petals and quite a few with eight. Daisies with 13, 21, 34, 55 or 89 petals are all quite common.

Lilies, irises, and the trillium have 3 petals; columbines, buttercups, larkspur, and wild rose have 5 petals; delphiniums, bloodroot, and cosmos have 8 petals; corn marigolds have 13; asters have 21; and daisies have 34, 55, or 89 petals—all Fibonacci numbers.

While this is certainly not a fixed rule and any member in a species can vary from the pattern, what is quite remarkable is the regularity with which the pattern persists.

Fibonacci Numbers and Phyllotaxis

The association of Fibonacci numbers and plants is not restricted to numbers of petals. It also shows up in the study of *phyllotaxis* (Greek for "leaf arrangement.") The phrase was coined by a Swiss naturalist named Charles Bonnet (1720–1793), who studied the patterns of seeds in a pinecone. He had noticed that their seed patterns were made up of opposite spinning spirals, and as he studied them he discovered that counting the number of parallel spirals in each direction will inevitably yield consecutive numbers along the Fibonacci series and that the ratio of clockwise to counterclockwise spirals is an approximation of Φ, the Divine Proportion. If the numbers are low in the sequence the approximation to Φ will be rough, and higher numbers produce a ratio that is closer to Φ. The graph that shows this is the same one we saw on page 128.

Pineapple

There are many, many examples in nature that exhibit this same structure, although they display different pairs of numbers in the sequence—sunflowers; pineapples; corn and wheat; needles of cacti; blackberries, raspberries, and strawberries; and the thorns of a rose are some examples, along with certain spotted patterns that appear on the tops of some seashells.

Strawberries

This phenomenon is not restricted to seed, thorn, and kernel patterns. It is even more common for spirals to wind around the cylindrical stem of a plant. Leonardo da Vinci once wrote, "The leaf always turns its upper side toward the sky so that it may be better able to receive the dew over its whole surface; and these leaves are arranged on the plants in such a way that one covers another as little as possible. This alternation provides open spaces through which the sun and air may penetrate. The

Fibonacci Numbers in Spirals

The florets in pineapples, sunflowers, daisies, and strawberries appear to form two systems of spirals, radiating from the center. Although these look like they are symmetrical, the numbers of clockwise and counterclockwise spirals are in fact not equal. When they are carefully counted they reveal a pattern of successive Fibonacci numbers.

The number of spirals in two directions may be 21 and 34, or they may be 8 and 13. The reason why this happens can be explained when we consider what happens during the early stages of the plant's life. If we look at the plant as the tip of a cone and consider the initial arrival of buds on the cone we will see that as new buds emerge the older ones are repelled outward and downward.

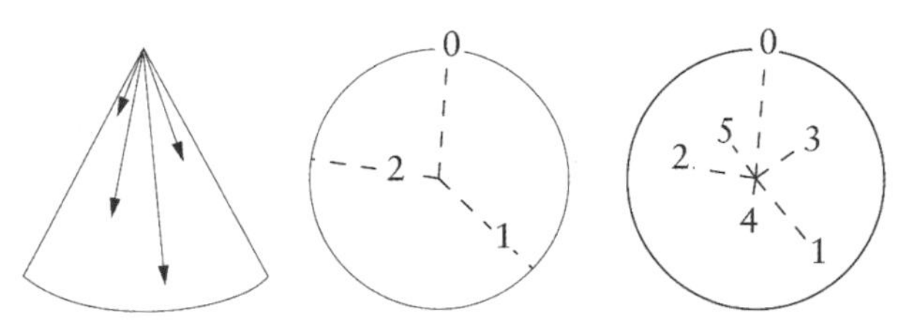

Numbering the buds 0, 1, 2, 3 ... in order of appearance shows what happens. Buds 0 and 1 separate the cone into larger and smaller sections. Bud 2 then finds it easiest to move into the larger section, forcing 3 into the smaller section. As the plant grows in this fashion the buds continue to grow into the largest space they can find developing into the spirals that we see.

arrangement is such that drops from the first leaf fall on the fourth leaf in some cases and on the sixth in others."

In many plants alternate leaves are distributed along the twig in a spiral pattern of growth. If we draw a line from the point of attachment of one leaf to the next, this line will wind around the twig as it rises. The same species will always bear the same number of leaves for each turn around the twig. An equal portion of the circumference of the stem will always separate the successive leaves from each other. The spiral shape created in this fashion is the equiangular spiral.

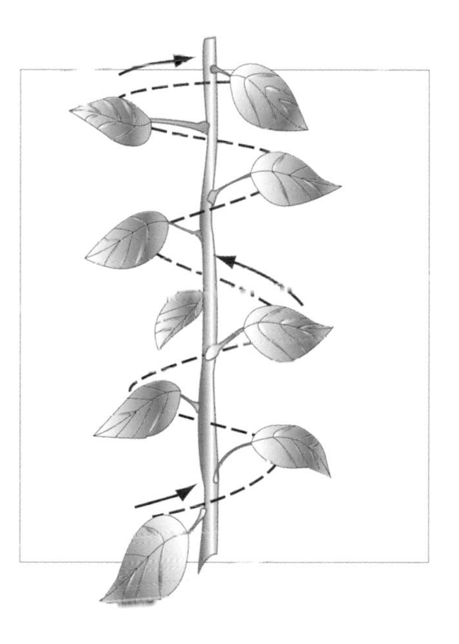

Ranking is determined by counting the number of leaves that occur in each 360 degrees of the spiral.

As the spiral progresses up a stem a certain number of leaves are formed before the spiral arrives back at a point directly above where the first leaf has emerged and creates another leaf. The relationship of the number of turns a spiral makes to the number of leaves that emerge can be written as a fraction and is called ranking.

If a line is drawn around the stem from the first leaf to the second and the angle between them is 180 degrees, then the third leaf will be found directly over the first leaf and the fourth will be directly over the second one. This is described as two-ranked and as the fraction 1/2. The numerator is the number of revolutions (1) around the twig and the denominator (2) means that two leaves were encountered in this one spiral of 360 degrees (not counting the first leaf). Examples of this are grasses, sycamore, birch, elm, and linden.

A further appearance of Fibonacci numbers is in the basis of all life, DNA. Although they are sometimes called the molecules of heredity, pieces of DNA are not single molecules but pairs of

More Spirals

The Fibonacci sequence of numbers, although appearing far too often to be discounted, is not accepted as a universal law in nature and scientists do not agree on a scientific explanation for phyllotaxis despite our fascination with the numbers involved. There is no doubt, however, that people down through the ages have observed and worshipped nature's spiral growth patterns and there are many works of art from very early times that display the spiral shapes of nature.

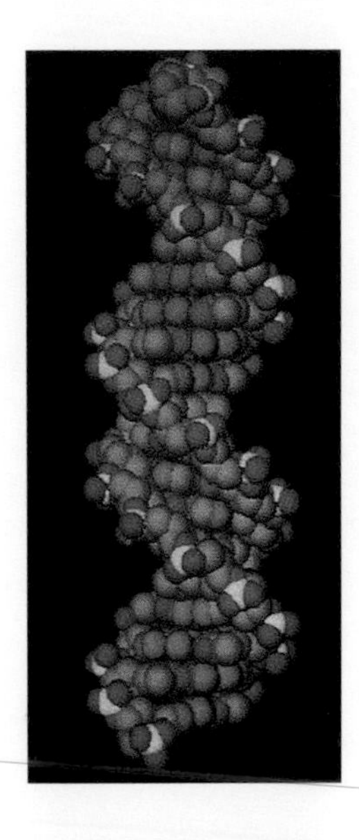

Spirals and DNA

The DNA molecule, the basis of all life, has some remarkable measurements.

It measures 34 angstroms long by 21 angstroms wide for each full cycle of its double helix spiral. Both 34 and 21 are numbers in the Fibonacci series and their ratio, 1.6190476... closely approximates Φ, which is 1.6180339....

The DNA in the cell appears as a double-stranded helix referred to as B-DNA. This form of DNA has two grooves in its spirals, with a ratio of Phi in the proportion of the major groove to the minor groove, or roughly 21 angstroms to 13 angstroms.

Furthermore, a cross-sectional view from the top of the DNA double helix forms a decagon—in which each spiral of the double helix traces out the shape of a pentagon.

Spirals and Art

Starry Night by Vincent van Gogh (1853–1890)

A section of a decorated ceiling from the tomb of Inherkau in Egypt (1140 B.C.E.)

molecules that entwine like vines to form a double helix. Each vine-like molecule is a strand of DNA: a chemically linked chain of nucleotides, each of which consists of a sugar, a phosphate, and one of four kinds of nucleobases (bases).

The cell's machinery is capable of disentangling a DNA double helix and using each DNA strand as a template for creating a new strand that is nearly identical to it. The errors that occur in this process are called mutations.

Nature, the mysterious forces that create the world we live in, is an experience of such beauty and awesome complexity that it is simply amazing to stumble across patterns that we can recognize and reduce to numbers and written rules. And yet it happens. As we peer deeply into the smallest parts of what makes up the whole, we find endless reflections of a proportion that describes so perfectly our place in it all—a perfect harmony of dynamics and balance, of known and unknown, of who we are and what possibilities we hold.

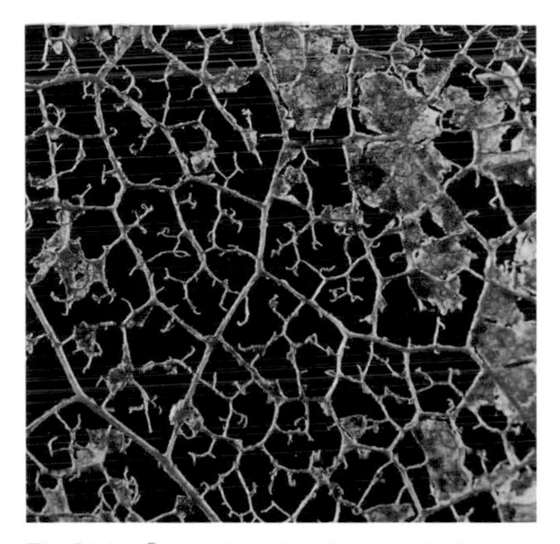

The Divine Proportion describes a ratio that compares the whole to the larger in exactly the same proportion as the larger is to the smaller. Nature describes this ratio in her own way. This picture of a dead magnolia leaf resembles a view of a town that might have been taken from the air. Just as our roads supply us with our own necessities, each vein in the leaf contains bundles of tubes which carry water and food to the cells of the living leaf.

The world, harmoniously confused,
Where order in variety we see,
And where, tho' all things differ,
All agree.

ALEXANDER POPE

Chapter 6

Divine Proportion in Science

The real mystery of life is not a problem to be solved, it is a reality to be experienced.

J. J. Van der Leeuw

Since ancient times, philosophers and scientists have searched for a way to describe the origin, structure, and order of the universe. With each new discovery, myths were dispelled and replaced by new "truths"—truths that often carried over into successive centuries, becoming firmly rooted in our collective consciousness. Over the past 2,500 years, there has been a series of individuals whose scientific discoveries or philosophical ideas have significantly changed the way we view the world and have given us new concepts to describe our place within the scheme of things.

In the 5th century B.C.E. Aristotle described methods of inquiry based on logic and reason. He devised a planetary system that he concluded must be sphere shaped and not flat because the shadow on the Moon during eclipses is always round. In this geocentric vision, Earth was stationary and the planets Mercury, Venus, Mars, Jupiter, and Saturn, as well as the Sun and the

An engraving from the book *Astronomiae* by Tycho Brahe (1546–1601) shows him standing at the mural quadrant at Uraniborg. Tycho used a huge mural quadrant, one of the largest astronomical instruments of its time, to make accurate measurements of the positions of stars and planets. He named his observatory Uraniborg, after Urania, the Muse of astronomy.

OPPOSITE: *The Sacrament of the Last Supper* by Salvador Dali (1904–1989)

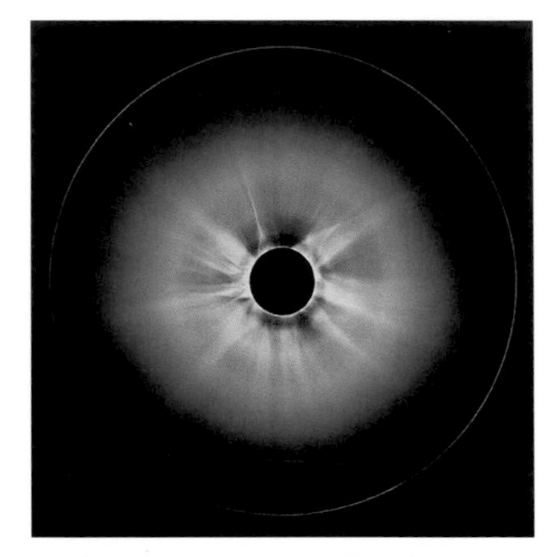

The Sun is the nearest star to Earth, lying ninety-three million miles away, and at the center of the solar system. Although we normally only see the bright disk, the Sun has an extensive, faint atmosphere called the corona, which extends millions of miles into space and is visible from Earth during an eclipse, as in this picture.

Tycho Brahe estimated the distance between the Sun and the Earth at five million miles. Later, Johannes Kepler (1571–1630) estimated the distance to be fifteen million miles. In 1672, Giovanni Cassini made a much better estimate of eighty-seven million miles, which is very close to the modern-day number.

Moon all encircled us. The stars were fixed to the celestial sphere that did not extend much farther than Saturn.

Aristotle devised a theory of a "prime mover"—a mystical force behind the fixed stars that caused the circular motions. In later times, with the advent of Christianity, the same theory was still adhered to and the mystical force was thought to be made up of angels.

This concept has, of course, changed many times. Earth is no longer at the center, and our universe is known to extend far, far beyond what our eyes can see. The stars are certainly not fixed to some celestial sphere. Furthermore, the Earth, according to the changing theories of scientists, has changed not only its shape but also the way it moves—many times.

In our contemporary times, physicists are again redefining the universe. Old methods of prediction and control are as outdated as the use of mathematical equations. The universe now appears to be an undivided wholeness enfolded into an infinite background source that unfolds into the visible, material, and temporal world of our everyday lives. We now have a universe where we recognize that thought can grasp the unfolded aspects, but only something beyond thought can experience the enfolded whole.

Astronomers Who Replaced Old "Truths" With New Ones

Ptolemy

Five centuries after Aristotle devised his planetary system, an Egyptian named Claudius Ptolemaeus (Ptolemy, 87–150 C.E.) created a slightly revised model that more accurately predicted the movements of spheres in the heaven, and helped account for observable planetary movements by setting Earth slightly off center of the universe. Western Christendom accepted this model, creating some space behind the fixed stars to accommodate a heaven and a hell, and the system reigned for well over a thousand years.

Portrait of Ptolemy from a 15th-century manuscript

Copernicus

In 1514 Nicolas Copernicus, afraid of being labeled a heretic, proposed a model for calculating planetary positions in which the Sun became the center of the universe. A priest himself, Copernicus was hesitant to divulge his theory that the Earth was just another planet in this heliocentric cosmos, and so withheld his work from all but a few astronomers until he was on his deathbed. He did not live to witness the chaos his theory would cause.

Nicolas Copernicus, a portrait from the town hall of Turin where he was born

Ptolemy's geocentric model of the universe

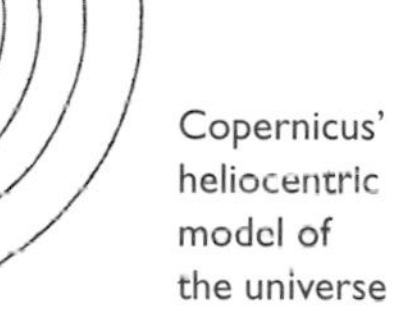

Copernicus' heliocentric model of the universe

ASTRONOMERS WHO REPLACED OLD "TRUTHS" WITH NEW ONES

GALILEO

In 1633 the Italian astronomer and mathematician Galileo Galilei (1564–1642) was taken to Rome to stand trial before the Inquisition for heresy. In defiance of an edict against the propagation of Copernicus' idea, he had forcefully asserted that it was the truth. Although he admitted to the Inquisition that he might have gone too far in his statements that the Earth moves around the Sun, the tribunal sentenced him to life imprisonment. Legend has it that as he rose to his feet, having sworn an abjuration on his knees, he muttered under his breath, "And yet, it moves."

Galileo at his trial

KEPLER

Like Copernicus, Johannes Kepler (1571–1630) was a deeply religious man. He believed that humans are made in the image of God and because of this are clearly capable of understanding the universe that God created. Moreover, Kepler was convinced that God made the universe according to a mathematical plan. He is remembered for his articulation of three major laws of planetary motion.

NEWTON

Isaac Newton (1642–1727) held the belief that simple and permanent mechanical laws govern the universe. He maintained that the fact that we are separate from nature enables us to observe the world objectively. Having set out to discover the cause of the planet's elliptical orbits, he deduced the laws of gravitation and then the laws of motion and matter. With a single set of laws he united the Earth with all that could be seen in the skies.

Cartoon of the story that Newton discovered gravity when he was struck on the head by a falling apple

KANT

Immanuel Kant (1724–1804) took God out of the scientific equation and viewed natural phenomena entirely as a mechanical scheme. According to Kant, the universe cannot be fully understood using sensory perception, thus a rational system like mathematics must be used.

EINSTEIN

Albert Einstein (1879–1955) proposed the theory of relativity and changed once again the concept of space, asserting that not only does a body's gravitational mass act on other bodies, it also influences the structure of space. If a body is massive enough, it induces space to curve around it. He sought to understand the mysteries of the cosmos by probing with his thoughts rather than relying on his senses. "The truth of theory is in your mind," he once said, "not in your eyes."

Einstein's discoveries led to the modern age in which a world picture is emerging that no longer reduces nature (and human nature) to a series of isolated parts. Modern scientists now view the cosmos as a web of interconnected and interrelated events. Theories of quantum physics show that humankind cannot use isolation as a method to gain knowledge of the whole.

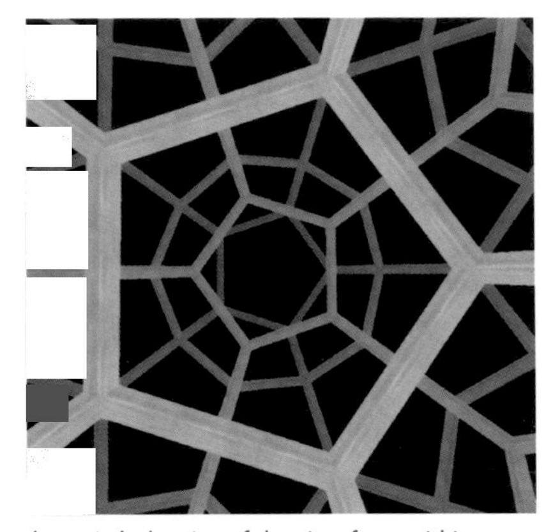

An artist's drawing of the view from within a dodecahedral-spaced universe

The Universe and Φ

In October 2001, NASA began collecting data on cosmic background radiation with the Wilkinson Microwave Anisotropy Probe (WMAP). This American satellite has been examining the microwave radiation generated shortly after the universe began—radiation that can tell scientists a lot about the physical nature of space. The wavelength of the radiation is remarkably pure, but like a musical note it has harmonics associated with it. These harmonics reflect the shape of the object in which the waves were generated. In the case of a note, that object would be a musical instrument. In the case of the microwave background, that object is the universe itself. In February of 2003, NASA released the first data from the probe, and in October a team of scientists used this data to develop a model for the shape of the universe.

Jean-Pierre Luminet and colleagues from the Observatoire de Paris used the information to conduct a study in which they analyzed a variety of different models, including flat, negatively curved (saddle-shaped), and positively curved (spherical) space. The study has revealed that if the Observatoire data is correct, it would predict a universe that is finite and is shaped like a dodecahedron (Greek for "two plus ten faces"). This is still a theory, but it is supported by data that can be tested. This closed universe should be about 30 billion light years across.

One of the astounding things about this discovery is its relation to Plato's proposition twenty-five hundred years earlier that the universe had finite limits. He also thought it was in the shape of a dodecahedron—one of his five so-called Platonic Solids.

Divine Proportion in the Dodecahedron

PLATO REPORTS IN *FIVE DIALOGUES* what the world looks like from above:

> *Well then, my friend, in the first place it is said that the earth, looked at from above, looks like those spherical balls made up of twelve pieces of leather....*

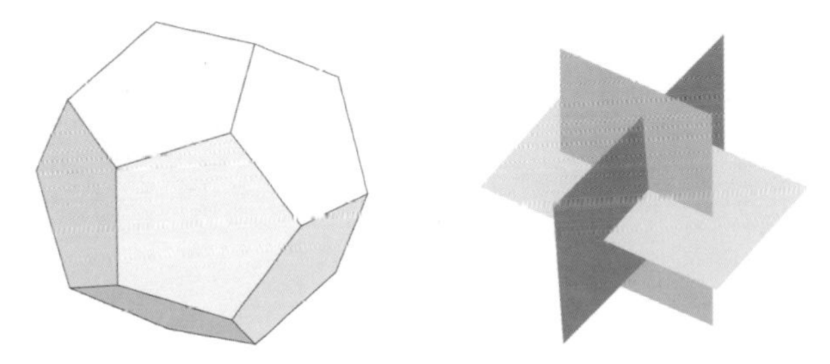

To help visualize the Divine Proportion in the dodecahedron space we see how this space can be constructed around three interlocking Golden Mean rectangles.

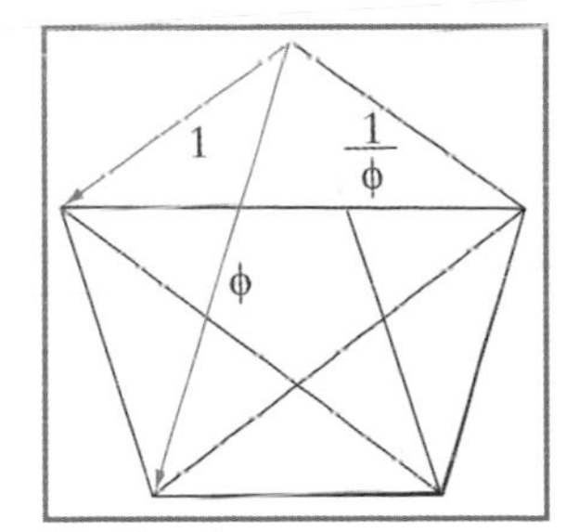

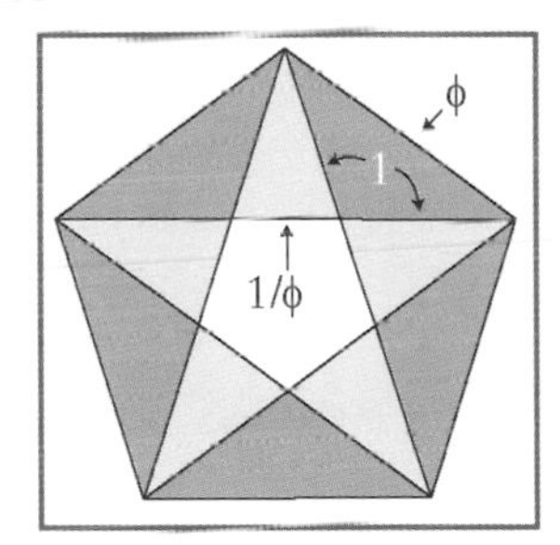

The connection of the dodecahedron to Φ is further revealed in the pentagons that form the twelve faces of a dodecahedron. Beginning with a pentagon, when lines are drawn to form a five-pointed star, the ratios of the lengths of the resulting line segments are all based on Φ.

The pomegranate has become a symbol for the dodecahedron and represents the fruitful and prolific nature of life that is always ready to burst forth and scatter its seeds.

The Greek legend of Earth's fertility explains that Persephone, daughter of Zeus, was abducted by Pluto, God of the Underworld, while she was out gathering flowers. Her mother, Demeter, sought her everywhere and threatened to destroy humankind by withdrawing fertility from Earth if she could not find her daughter. Zeus promised to restore Persephone to her mother provided Persephone had eaten nothing in the underworld; but she had eaten some pomegranate seeds. She was therefore compelled to spend six months each year with Pluto and then allowed six months with her mother.

The Platonic Solids

The infinite! No other question has ever moved so profoundly the spirit of man.
David Hilbert (1862–1943),
German mathematician

Of the many mystical geometric concepts developed by the early Greek philosophers and related to the Divine Proportion, this is one that has run alongside the achievements of science in a very interesting way. The series of shapes called the Platonic Solids—called so because of Plato's description of them in *Timaeus*—are more formally known as the "regular polyhedra." Each one of the five forms fits perfectly within a sphere presenting an identical view in all directions and their surfaces all have the same shape. There are only five volumes that fulfill the requirements of equality by repeating identical corner angles, edge lengths, and surface shapes within a sphere.

The names of the five Platonic Solids derive from the number of faces they have. Each was conceived as representing one of the elements or states of matter and the fifth was seen to represent the fifth element or cosmos that held them all together.

Cosmos	Dodecahedron	Pentagon	12
Fire	Tetrahedron	Triangle	4
Air	Octahedron	Triangle	8
Water	Icosahedron	Triangle	20
Earth	Hexahedron	Square	6

These five shapes were looked upon with a sense of awe in ancient times. The construction and study of their forms were considered the ultimate goal toward which those who studied numbers would arrive and represented a pinnacle of ancient geometric and esoteric knowledge. Their construction comprised the final books of Euclid's *Elements*.

Plato and the Platonic Solids

PLATO (CA. 427–347 B.C.E.)

Plato was born to an aristocratic family in Athens. His real name was Aristocles and the nickname "Plato" apparently originated from wrestling circles. The word Plato means "broad," and probably refers either to his physical appearance or to his wrestling style. As a young man Plato had political ambitions, but he became disillusioned by the political leadership in Athens and eventually became a disciple of Socrates, accepting his basic philosophy and dialectical style of debate. It is mostly through the writings of Plato that we have details of his great master.

In 387 B.C.E. Plato founded an Academy in Athens, often described as the first university. It provided a comprehensive curriculum that included astronomy, biology, mathematics, political theory, and philosophy. His final years were spent lecturing at his Academy and writing. He died at about the age of 80 in Athens.

Timaeus and *Critias*, composed around 360 B.C.E., are two of Plato's dialogues that are written as conversations between Socrates, Timeaus, Critias, and others, apparently in response to a talk Socrates gave about ideal societies.

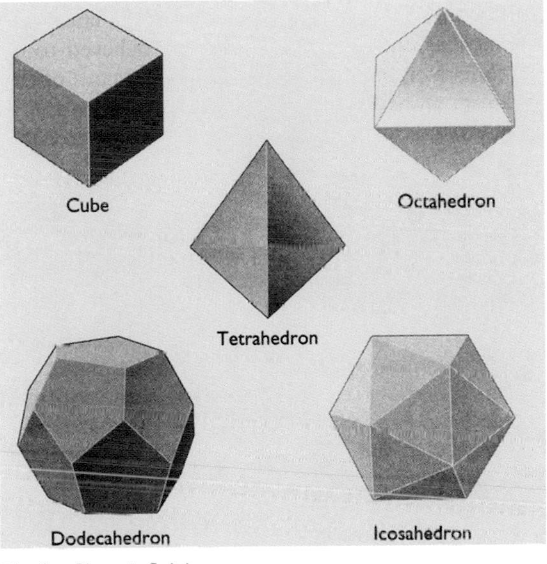

The five Platonic Solids

In the dialogue *Timaeus*, Socrates tells a story, which Plato professes through Socrates' voice to be true, relating a conflict between the ancient Athenians and the Atlantians 9,000 years before Plato's time. This knowledge of the distant past had apparently been lost to the Athenians of Plato's day, but the story of Atlantis had earlier been conveyed to Solon by Egyptian priests and Solon had passed the tale to Dropides, the great-grandfather of *Critias*. Timaeus, a Pythagorean, philosopher, scientist, and contemporary of Plato, does most of the talking and describes the geometric creation of the world.

Mosaic from Pompeii (1st century C.E.) called *Conversation Between Philosophers* (also known as *The School of Plato*)

Trying to create a mental image of some scientific discoveries is as difficult as imagining a four-dimensional object from our three-dimensional view. Plato originally made this point in his allegory of the cave in *The Republic*. A race of people has been constrained from birth to live in a cave. All they know of the outside world is the information they can glean from the colorless shadows that are cast on the walls of the cave.

The idea that all things are composed of four primal elements, earth, air, fire, and water, is attributed to Empedocles (ca. 493–433 B.C.E.), a Greek philosopher and poet who was a disciple of Pythagoras.

In *Timaeus*, Plato put forward the suggestion that the four elements were all aggregates of tiny solids (in our modern parlance, atoms) and that since the world can only have been made from perfect elements, they must be derived from the shapes of the five regular solids.

First, he argued, we must have fire, to make the world visible, and earth to make it resistant to touch. As the lightest of the elements, fire must be a tetrahedron, he argued, and as the most stable, earth must consist of cubes. Being the most mobile and fluid, water was assigned to the icosahedron, the one most likely to roll easily. Air, Plato observed, is to water as water is to earth and must be an octahedron. Finally he assigned to the entire universe the shape of the dodecahedron.

Although Plato's theory now appears to be rather whimsical and fanciful, the idea that the regular solids played a fundamental role in the structure of the universe was still taken seriously in the 16th and 17th centuries when Kepler began his quest for mathematical order.

There were six known planets in Kepler's time: Mercury, Venus, Earth, Mars, Jupiter, and Saturn. Influenced by Copernicus's theory that the planets move around the Sun, Kepler tried to find numeric relations to explain why there are exactly six planets, and why they are at their particular distances from the Sun.

KEPLER AND HIS THEORY OF THE MUSIC OF THE SPHERES

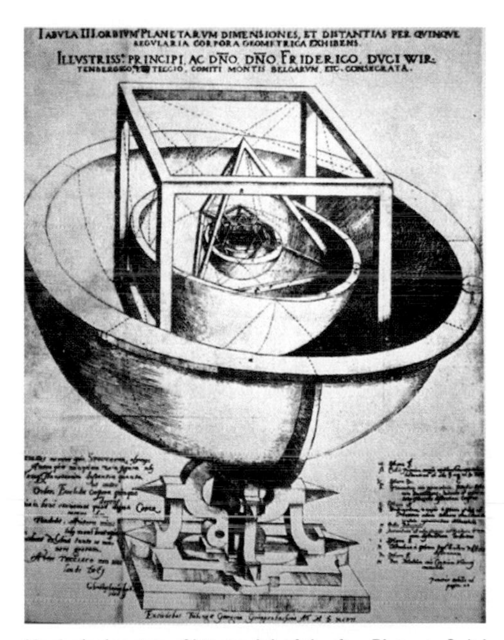

Kepler's drawing of his model of the five Platonic Solids to illustrate his theory; unfortunately it is all wrong.

IN HIS 1609 COMMENTARIES *on the Movements of the Planet Mars* Kepler proved the first two of his laws of planetary motion: that planets move not in circles but in ellipses; and that planets sweep out equal arcs in equal times. Then, in his 1619 *Harmonies of the World,* he proved the third: that the cube of the planet's mean distance is equal to the square of its time of rotation.

It was in this latter work that he republished his own wonderful speculation that Euclid's five regular solids can be inscribed within spheres, very nearly intersecting the orbits of the planets. He had first published this view in 1596, when he still thought, with Copernicus, that the planets moved in circles and hence the solids could be inscribed within the spheres.

Kepler found that the diameters of the planets' orbits are in small whole number ratios to one another, just like the tones in musical scales: 1:2, 2:3, 3:4, etc. Mars, for instance, is about half as far from the Sun as Jupiter, that is, as 1:2, or an "octave" higher. Kepler worked out many such small whole number relationships, concluding that the planets move to the "music of the spheres." He interpreted his discovery as revealing the mathematical methods of God the "very wise Founder" of the heavens.

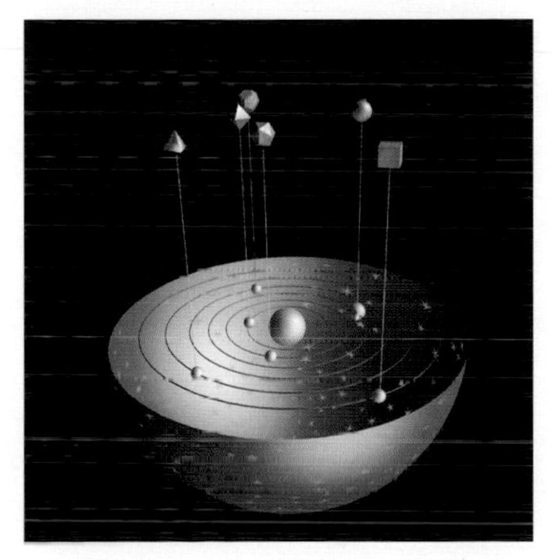

Kepler's view of the universe linked the planets with the Platonic Solids: Mars as dodecahedron, Venus as icosahedron, Earth as sphere, Jupiter as tetrahedron, Mercury as octahedron, Saturn as cube.

A cross-sectional view from the top of the DNA double helix forms a decagon, which is in essence two pentagons, with one rotated 36 degrees from the other, so each spiral of the double helix must trace out the shape of a pentagon.

Kepler decided eventually that the key is not numeric, but geometric. There are six planets, he reasoned, because the distance between each adjacent pair must be connected with a particular regular solid, of which there are just five. After some experimentation, he found an arrangement of nested regular solids and spheres, so that each of the six planets had an orbit on one of the six spheres.

Although it is hard for us to imagine that two intellectual giants of the caliber of Plato and Kepler should have proposed theories so unfounded on scientific data, they were clearly driven by the same deep-seated belief that motivates today's scientists—that the pattern and order of the universe can all be described and to some extent explained by mathematics.

And however absurd Plato's convictions actually were, the variations that nature has made on the five solids are virtually endless. As the skills of mathematicians developed over the following centuries, a multitude of fascinating properties began to appear. These properties apply to all crystals, the orderly repeated arrangement of atoms, and the folding of proteins in our DNA.

I think that modern physics has definitely decided in favor of Plato. In fact the smallest units of matter are not physical objects in the ordinary sense; they are forms, ideas which can be expressed unambiguously only in mathematical language.

WERNER HEISENBERG (1901–1976),
GERMAN PHYSICIST

Φ AND THE SCIENCE OF CRYSTALLOGRAPHY

We are all aware of the beautiful patterns that nature creates and because they so accurately describe the principles that provide scientific solutions to all sorts of puzzling situations, they have long been studied. Designs in nature that were at first contemplated out of curiosity have had surprising implications in our modern world. An example of this concerns the branch of science called crystallography, which deals with the geometric description of crystals and their internal arrangement. Crystallography is relevant to our world because an accurate understanding of molecular structures is a prerequisite for drug design. What has been discovered in this field—a discovery that relates to the Divine Proportion—is as significant as the discovery of another color in the rainbow would be to our experience of color.

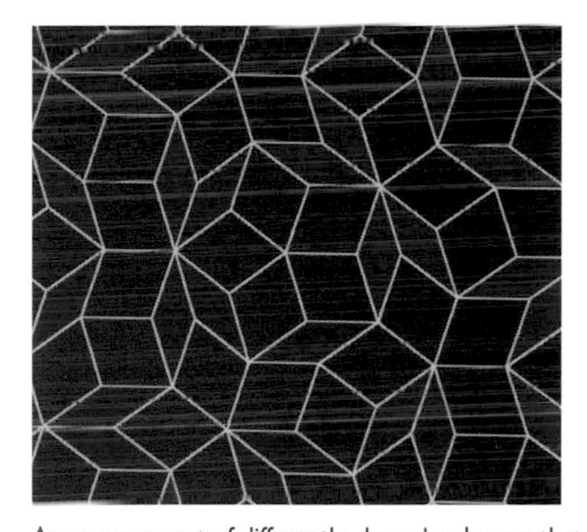

An arrangement of differently shaped polygons that completely cover a plane with no gaps or overlaps is called "tiling the plane." These arrangements can be periodic or aperiodic. This aperiodic arrangement is made up of only two polygon shapes and is an example of Penrose tiling—an arrangement that forms the basis of our understanding of substances called quasicrystals.

In 1984, materials engineer Dany Schectman and his collaborators found that certain crystals display a property that researchers had previously deemed to be only theoretically possible. However, they had seen no proof of it and therefore could not substantiate the theory. To understand the nature of this discovery it will help to look initially at two related principles, those of packing and of tiling.

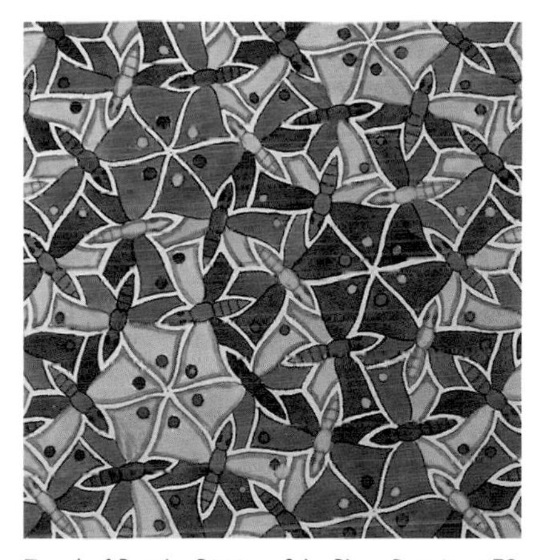

Detail of *Regular Division of the Plane Drawing #79* by M. C. Escher is an example of periodic tiling.

As we saw in the previous chapter, nature employs many principles that provide an optimal situation for growth. Specifically, we looked at the Fibonacci numbers and the phyllotaxis of plants. There we saw some interesting appearances of the Divine Proportion especially as it pertains to spirals and the Fibonacci numbers. In the early part of the 17th century, before Charles Bonnet had begun his research into the subject, Johannes Kepler

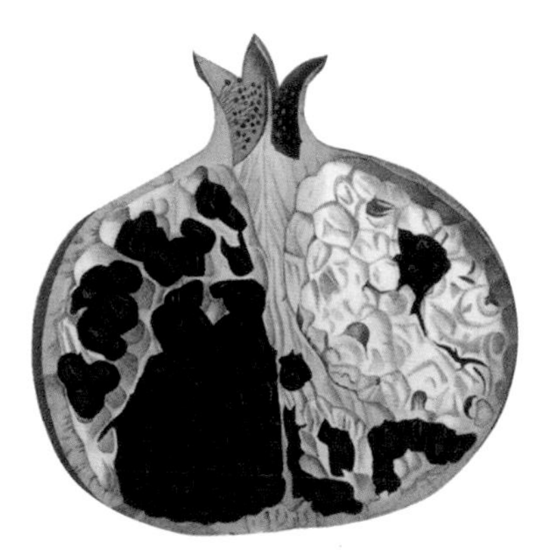

Pomegranate seeds grow into dodecahedrons.

had also studied it. Fascinated by the way that seeds are stacked in a pomegranate he looked a little deeper. His observations led him deep into the enormous symmetry of nature. He eventually determined that matter in its solid state can be of only two basic forms—highly ordered or amorphous.

The molecules of a solid do not move around very much. Compared to the molecules of liquids and gases they tend to stay put relative to each other. If a solid has molecules arranged in an orderly fashion, we say it is ordered, or *crystalline* (ice, diamonds, table salt, and sugar). If the molecules of a solid are not arranged in any order, we call the solid *amorphous* (ceramics, some plastics, and glass).

A lattice will help determine if a pattern is regular or not.

Of these two states, that of crystals is thought to follow several rules. One is that the units of crystals are periodic, meaning that they have a repeating lattice structure with a definable unit cell that can be stacked on itself and repeated to create the whole crystal. The second rule is that this lattice can only contain certain kinds of symmetry. Two-, three-, four-, and six-fold symmetry is allowed. Five-fold symmetry is forbidden because a solid cannot be packed with regular five-sided shapes without leaving holes or spaces.

The Science of Packing

When Kepler looked closely at the packing of seeds in the pomegranate, one of the first things he did was compare it with other similar structures, particularly those of snowflakes and honeycombs. All three forms are designed around geometric principles of regularity.

Assuming that nature always adopts the most efficient means to achieve her ends, the regular patterns of pomegranates, honeycombs, and snowflakes can all be explained by examining a system of spheres and observing the geometric solids they give rise to when they are packed into a three-dimensional space. For example, when spheres are packed in a cube they will all expand into cubes, and spheres arranged in a hexagonal space will all expand into hexagonal prisms.

In snowflakes (above) and honeycombs (below) Kepler observed the principles of packing.

The seeds of the pomegranate are initially tiny spheres. As they grow and expand fully into the shape the skin of the pomegranate will allow, they finally grow into the form of the twelve-sided dodecahedron we have already looked at, poking the pomegranate's soft skin into little bumps and making optimal use of the space inside. Turning to honeycombs and snowflakes, Kepler saw the same dynamic at work.

Although this was all very interesting to him, Kepler knew he was just beginning to understand the principles at work in the phenomena in the world around him. Later researchers discovered that there are only four three-dimensional shapes that pack perfectly. Two-, three-, four-, and six-sided shapes are the only ones that will pack a space with regularity and with no gaps.

While the rules that apply to packing in three dimensions were quite easy to deduce, the problem that occurs in two-dimensional space is more complicated. To understand it more completely mathematicians turned to what is called tiling. Tiling asks the question: What shapes can be stacked together to fill a two-dimensional space (or plane) completely?

PACKING

NATURE DISPLAYS THE ART OF PACKING in instances that have prompted scientists to look deeply into the mathematical principles involved to discover secrets of efficiency. Regular five-sided shapes cannot pack without leaving unused space. Regular shapes with two, three, four, and six sides will pack most efficiently.

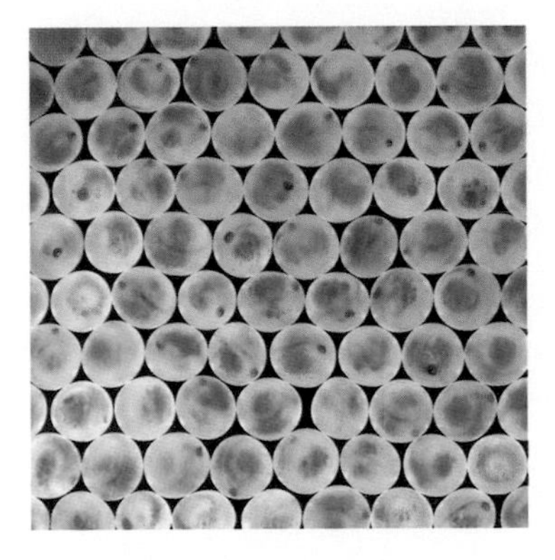

This clutch of perfectly packed rainbow trout eggs is an example of nature's technique of packing spheres. The female trout lays up to 5,000 eggs in a nest on the gravelly riverbed. We copy this technique when we pack oranges in boxes.

In 1915, using the newly developed techniques of the X-ray, it was demonstrated that snowflakes consist of identical particles arranged in regular lattices. The snowflake starts out as a tiny, hexagonal crystal seed of ice in the upper atmosphere. As air currents carry the seed up and down through different altitudes the crystal grows. The pattern that evolves depends upon the seed's particular movements, but since the snowflake is so small, the same pattern of growth occurs on all sides, and hence the hexagonal shape of the original seed crystal is preserved.

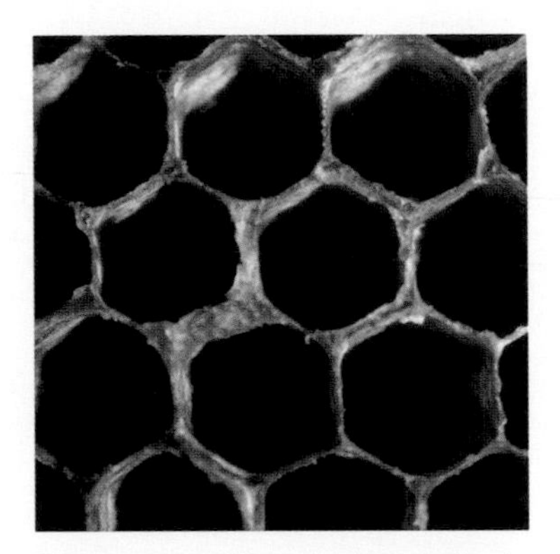

There were initially many theories developed to describe how the hexagonal wax shape of the honeycomb is created. Charles Darwin thought that bees first hollow out cylinders of wax and then push out the walls of each cylinder until they hit the neighboring cells, filling up the empty space between them. What we now know is that the bees secrete the wax as solid flakes and construct the honeycomb cell by cell, face by face—a highly skilled geometer, equipped by nature for the task of constructing its honeycomb in the form that is mathematically the most efficient.

The Science of Tiling

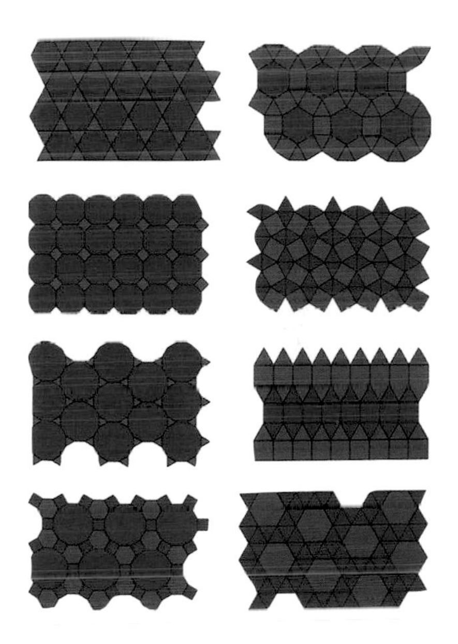

The eight types of symmetric tiling that are possible using regular shapes The restriction is that the patterns must be similarly arranged around each vertex.

In the natural world tile-like phenomena are evident everywhere—on a turtle's shell, the scales of a fish, and the cells of our skin. Over the centuries artists all across the globe have used tiles to tessellate floors, paintings, and walls in creative and really beautiful ways.

In a two-dimensional plane, there are three—and only three—regular shapes that can be arranged to completely fill, or tile, the plane. These are the equilateral triangle, the square, and the regular hexagon.

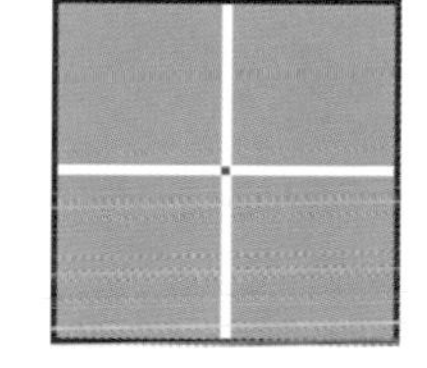

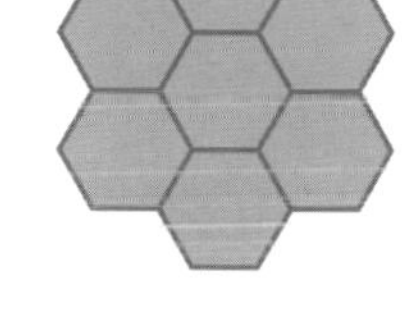

Pentagons will not work.

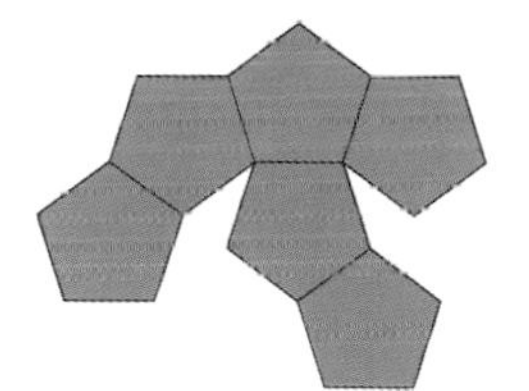

If two or more regular shapes are allowed, with the restriction that the same array of shapes surround each vertex, then there are exactly eight further possibilities, made up of combinations of triangles, squares, hexagons, octagons, and dodecagons. These eight possibilities were discovered by Kepler.

Tiling

Tiling is an art. Using regular and irregular shapes nature's patterns have inspired artists to create extremely complex and beautiful designs of their own.

Tiling on a turtle's back

The scales of a fish are another of nature's examples of tiling.

Chaos is present everywhere in countless ways and forms, while Order remains an unattainable ideal.
M.C. Escher

Tiling using a mixture of regular and irregular shapes has been practiced around the world.

Friday Mosque of Isfahan

M. C. Escher has created ingenious tilings of objects including many that appear to be in motion through three dimensions.

If nonregular shapes are allowed, then there is no limit to the number of possibilities of regular tiling. In particular, any triangle or any quadrilateral can be used, but only those pentagons will work that have a pair of parallel sides. In the case of hexagons, it has been proved that there are precisely three kinds of hexagons that will tile a plane with regularity. Beyond these shapes there are no polygons (with seven or more sides) that can be arranged to completely fill the plane in a repeating pattern.

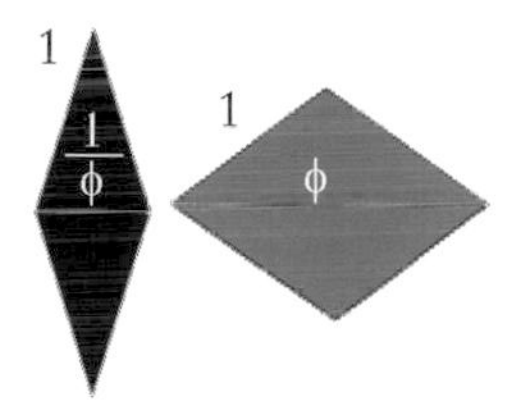

These tiles can be combined in various ways. As the patterns are expanded to cover greater areas, the ratio of the quantity of the one type of tile to the other always approaches Φ, or 1.6180339..., the Divine Proportion

Just as the solid states of matter fall into two types, tiling also can be separated into two types. One works with repeating (or *periodic*) patterns and has translational (repeating) symmetry. The other comprises tiling with no translational symmetry, and is called *aperiodic* tiling. In 1964 an aperiodic set of tiles was discovered that consisted of 20,000 differently shaped tiles that fit together perfectly. After this discovery many designs were created that used far fewer tiles.

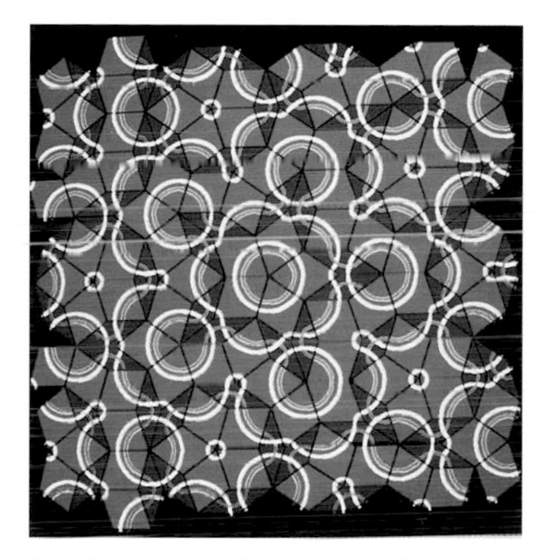

Part of an (aperiodic) Penrose tiling of the plane exhibiting the local fivefold symmetry of the tiling

While quite a lot is known about periodic tiling, aperiodic tiling is still being studied. In 1974 the British mathematician Roger Penrose discovered a single pair of polygons that can be fitted together to completely fill a plane in an aperiodic fashion—meaning the tiles fit together perfectly but there is no symmetry. In order to force the aperiodicity, the edges of the tiles were given certain restrictions in the way they could be aligned.

One of the very interesting things about these shapes is that they are related by the ratio of Φ.

Further, when a plane is tiled with these figures, the ratio of the number of fat tiles to thin tiles is the ratio of Φ or the Divine Proportion.

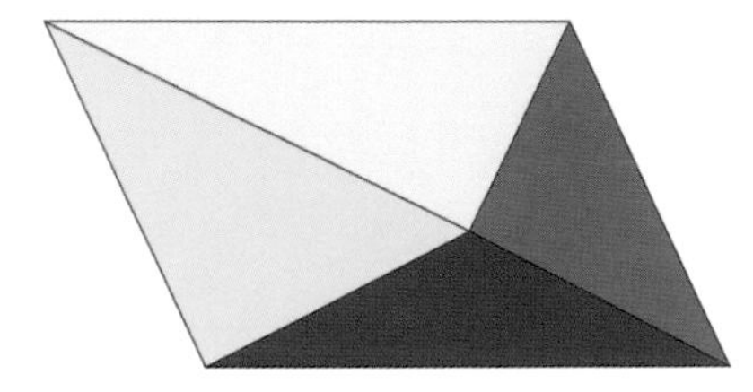

Six-sided, diamond-faced shape whose diagonals are in the ratio of Divine Proportion

Φ and a New Discovery

This same tiling pattern also extends into three dimensions with a single shape derived from the Penrose tiles. Just as it had long been thought that tiling a plane with five-fold symmetry is impossible, and that filling a three-dimensional space in five-fold symmetry was impossible; a semisolution has been found for two dimensions that requires two shapes, and in three dimensions one has been found that uses just one shape. The shape has six sides, each one a diamond whose diagonals are in the ratio of the Divine Proportion.

At this point, what seems to look like a mathematical exercise becomes the doorway through which scientists peer into a world they do not completely understand. Dany Schectman (the crystallographer) made the interesting discovery that the crystals of an aluminum manganese alloy exhibit the same principles as the semisolution to packing and tiling—principles that depend upon shapes of Divine Proportion. This was a stunning revelation to crystallographers. It appears that there is a set of crystals (i.e. a type of matter)—called quasicrystals—that are neither periodic nor amorphous. Remember that scientists have previously thought matter in its solid state can be of only two basic forms, highly ordered or amorphous.

What appears to happen in the alloy can be explained by a decagon-shaped tile discovered by the German mathematician Petra Gummelt. When it is inscribed with a particular pattern, which is again closely related to Φ, it will overlap in such a way as to tile a plane under a peculiar new rule that involves sharing. This suggests that when shaded areas of the tile are allowed to overlap, the plane can be tiled with the decagon shape.

When a side of this decagon is one unit, the radius of the circle that circumscribes it is equal to Φ. Interestingly, this is the same shape we saw in the cross section of DNA.

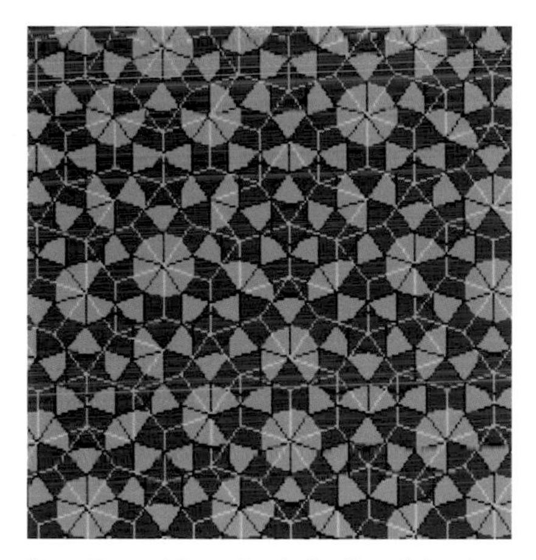

Petra Gummelt's quasiperiodic tiling of the plane with overlapping decagons and the built-in relationship to Φ

The principle offered by the decagon of Petra Gummelt has suggested to scientists what might be happening in quasicrystals: They may be clusters of atoms sharing atoms with neighboring clusters. If this proves to be the case, what we will have is a principle that describes a solid that is more—rather than less—stable.

Whatever implications this has for the world of crystallographers, it has wonderful implications in the world of human dynamics, for it gives a scientific example in which sharing creates a situation that has more inherent strength than the situation that lacks sharing.

While all of this continues to be investigated, scientists, stimulated by their curiosity and their love of probing into mysteries of our universe, continue to push the edges of the world we think we live in. With great thanks to them for the time they devote to the task, we must continue to remember that it might be somewhere in the overlapping aspects of our existence that we will find what we look for—somewhere in an understanding that includes more or is greater than the particular parts. The code—if there is one—that describes everything, will surely emerge from some wrinkle of time where life is experienced as the divine sharing of physical and mystical forces.

The most beautiful thing we can experience is the mysterious. It is the source of all true art and science.

ALBERT EINSTEIN

Chapter 7

Divine Proportion in Mysticism

Perfect comes from perfect. Take perfect from perfect, the remainder is perfect.
ISHA UPANISHAD

Detail from *The Last Judgement* (ca. 1320) in the Byzantine church of Kariye Camii. A throng of angels fills the background, and above them a single angel carries the scroll of the heavens.

ALTHOUGH MANY ATTRIBUTES OF Divine Proportion are not obvious, they are also not hidden; they simply need to be teased out. Implied in its mathematical expression $\Phi = (1 + \sqrt{5})/2$ are forms that carry a deep symbolism and patterns that express harmonious beauty. Carried into nature, its dynamism propels a regenerating spiral; in science its form helps define new dimensions. It represents a significant relationship of parts to the whole, and concealed within its encoded teachings are laws of sacred creativity, and hints of an experience that transcends the serious world of restrained thoughtfulness and thrusts its participants into an uncharted journey toward the ultimate reward of self-revelation.

There is, or was, a secret school (a brotherhood or a monastery) that existed in Afghanistan for thousands of years called the Sarmoun Darq. The name of the school means "beehive" or "collectors of honey." G. I. Gurdjieff, one of the world's great

OPPOSITE: *Sudama Approaching the Golden City of Krishna* by an anonymous Indian artist (ca. 1785)

Karnak Temple, Egypt

One of the most ancient of all mystery schools, the temple complex at Karnak was built to reflect the shape of a human being.

spiritual teachers, is said to have searched for many years for the source of esoteric teachings and to have finally found what he was looking for in the Sarmoun Darq.

The reason for this school—and there have been and still are many such schools—is to gather and preserve certain kinds of knowledge regarding the human soul, particularly at times when such knowledge appears to be dissipating. This activity and its significance have a profound implication for our human evolution. Without it we cease to evolve.

Mystics and masters collect and store sacred knowledge and esoteric principles in much the same way as bees collect nectar, although they have the capacity to concentrate and change the nectars when they are gathered so that when the time comes and the containers are accessed, the knowledge that has been concealed will inform the deeply curious and be of value to seekers of truth. By bringing new meaning to old texts, or revising rituals and meditation techniques, great teachers create new bodies of work. Secrets of esoteric understanding are redesigned and couched in new forms. Methods are borrowed from one tradition and used to inform another. Legends are created and masterfully designed nuggets of wisdom are hidden within. As time passes, the truths that once were obvious become obscured. They are hidden, and they wait to be rediscovered.

There are wonderful fragments of sacred knowledge in myths and sacred rites, in sacred sounds, in sacred art and architecture, and there are profound messages in the holy scriptures from every culture that have been handed down through the ages.

Many cultures tell stories of journeys and mythical quests in which a hero or heroine, faced with a crisis, and called upon to aid themselves or the cause of humanity, sets forth. Resolution is found through beautiful symbolism, and the seekers return home to find peace and deeper harmony. Filled with wisdom and cleverly disguised teachings, these tales are told and retold and become embedded in the minds and hearts of listeners. When real meaning becomes obscured by time, as sometimes happens, and false understandings creep in, truths simply lie dormant. When the time is ripe and the original meanings become apparent again, the tales take on a renewed life as wisdom teachings.

Many cultures build temples based on ideas of space that represent the cosmos in miniature or some aspect of the creation myth in which the divine is made manifest. Cathedrals, mosques, synagogues, and churches all incorporate structural design elements that are intended to intensify feelings of devotion. Many people experience what they describe as a divine state when they walk into these sacred spaces, and this is not accidental. Accoustics, colors, light, and many hidden elements contribute to an experience that has been refined by centuries of experimentation. For long periods of time the designs are carefully defined by religious law, then time passes and the laws are abandoned. People forget how to build their sacred spaces, and were it not for their permanence, very specialized knowledge would be lost. But the keys are there, locked into stone and embedded in shape.

The spiral shape of the Mosque of Samarra (9th century) in Iraq is symbolic of the holy mountain, the expansion and evolution of consciousness, and the ascent toward the flame of wisdom.

Dante's Inferno by Sandro Botticelli (1446–1510) shows a spiral cross section of the pit of hell.

SYMBOLS OF PERFECTION AND DIVINE KNOWLEDGE

The sacred Eye of Horus is an Egyptian symbol of fertility, and of death and rebirth onto a higher winding of the spiral journey of life.

This stone tablet from Babylon is in the shape of a Golden Rectangle and depicts the divine relationship between humans and God.

The hand gestures, called *mudras*, that we see in Buddhist sculpture and painting are generally used to indicate the nature of particular aspects of Buddha-nature. The gestures themselves have been developed over thousands of years as a symbolic language conveying complex religious concepts.

This opening page from the 7th century Irish Book of Durrow was designed to be used as a meditation to prepare readers for the truth of the Gospels that follow. Made up of one single line, these Celtic spirals represent the continuous creation and dissolution of the world.

Sacred Shapes of Divine Proportion

In the course of unraveling the Divine Proportion we have seen several shapes emerge—Golden Spirals, Golden Rectangles, Golden Triangles, and the Sacred Pentagons—all showing themselves in manifestations of art, nature, and science in a rich tapestry of mystical meaning and divine laws of proportion.

The Divine Proportion is a universal symbol for perfection and beauty. When it appears in any of its shapes in a work of sacred art, we are given a clear invitation to experience a higher harmony. Standing before such sacred art we feel a mysterious and immediate presence. Many feel changed by it, although they are not sure how. Many are drawn to it, but do not know why. These are the specific intentions of those who create such art. Symbols, sounds, and patterns that are deeply embedded in the human psyche are activated and those who feel their charge are, to some small degree, transformed.

Chartres Cathedral, built in the form of a cross, is oriented to the four quarters, with its altar facing east and with its many details controlled and inspired by laws of proportion. On an outer wall is a statue of Christ holding the Bible. The bottom corner of the Bible touches Christ's lower diaphragm, and the top corner touches his heart. The Bible itself is in the shape of a Golden Rectangle and bridges the centers of the subtle body.

It is not magic or guesswork that creates sacred art. It is the result of deep understanding and rigid discipline. Works of sacred art, architecture, music and poetry are among the highest achievements of human consciousness and they are designed with the specific intention to raise levels of awareness in those who stand in appreciation. They touch the subtle centers of the body and open the heart to something "otherworldly." They use something intangible, called beauty, to create a moment of awe.

Beauty is something we all recognize: It is something that cannot be defined but is transcendental, timeless, and immortal. Beauty is an experience of feeling. It simply cannot be properly

The Cathedral of Chartres was built between 1194 and 1260. The architect is unknown, but the span of its nave was the largest that had been done at the time. No one knows how the builders calculated the span, no one knows why they believed it would work—but the design of the interior is based precisely on a five-pointed star and embodies the Divine Proportion.

described. We see it, we feel it, we hear it, we know it; but we are at a loss when trying to adequately put it into words.

Perfectly proportioned architecture is of course one of the Divine Proportion's greatest and most powerful expressions in sacred art. The Parthenon in its present state of ruined glory still gives us a message that all but those most blind can see and feel. The cathedrals of the Middle Ages express a reverence for the laws of proportion that is all but forgotten in buildings of today. The society that built them was fulfilled by their creation. Their stones tell a story of time, effort, and love that was poured into creating something of real permanence.

Divine Proportion and Wholeness

The Divine Proportion is an expression of our relation to wholeness. One of the great paradoxes of Western tradition is a belief in the oneness of God and the unity of existence—and yet the multiplicity in the world we experience. The problematic result of this is the awkward belief that we are somehow separate from the rest of life.

Before Aristotle separated things into systems of cause and effect, Greek mystical philosophers such as Pythagoras, Heraclitus, Socrates, and his disciple Plato had spoken of a universe in which all things are one.

> *One thing arises from all things, and all things arise from one thing.*
>
> Heraclitus, Fragment 10

The words of great teachers go straight to the point, but when reading Heraclitus, whose teachings have been so badly preserved, it is often easy to miss the point or misinterpret, for Heraclitus has not left us a mathematical theorem, as Euclid did, but an existential truth. A precursor to Euclid's expression of Divine Proportion through a logical progression of thought, Heraclitus has expressed the principle as a truth he has known within himself.

The Death of Socrates by Jacques-Louis David (1748–1825)

Socrates (469–399) left no writings behind and is best known today through his appearance in the *Dialogues of Plato*. He was a widely recognized and very controversial figure in his native Athens, so much so that he was frequently mocked in the plays of dramatists. Plato, however, portrayed him as a man of great insight, integrity, self-mastery, and argumentative skill. At the age of seventy Socrates was brought to trial on a charge of impiety and sentenced to death by poisoning (probably hemlock) by a jury of his fellow citizens. Plato wrote *Apology*, which purports to be the speech Socrates gave at his trial in response to the accusations that had been brought against him (Greek *apologia* means "defense.") In this speech Socrates speaks to the necessity of doing what one thinks is right, even in the face of universal opposition, and of the need to pursue knowledge even when opposed.

Heraclitus

Heraclitus (late 6th century B.C.E.) is the most significant mystic of ancient Greece until Socrates and his disciple Plato, and his teachings are perhaps even more fundamental to the formation of the Western mind than all others, for he proposed a model of nature and the universe that created the foundation for all further speculation on the nature of things. He taught that everything is constantly changing and that there is an underlying *Logos* (proportion) to it all.

Very little remains of his writings. All we have are a few fragments, quoted by other Greek writers, and they give us only a taste of who he was and what he taught. These passages are tremendously difficult to read, not merely because they have no context, but because Heraclitus apparently quite deliberately cultivated an obscure writing style—so obscure, in fact, that the Greeks nicknamed him *ho skoteinos*, the obscure, the dark, the riddling.

Heraclitus lived in Ephesus, an important city on the Ionian coast of Asia Minor, and he was, by all accounts, not a very sociable creature. Diogenes Laertius reports that Heraclitus

Diagram of the Supreme Ultimate from the Taoist *Compendium of Diagrams* (1527–1608)

The Tao that can be spoken is not the true Tao.
Any name which can be given to it is not its true name.
That which was before Heaven and Earth is Tao.
Tao is the mother of all things.
The wise seek its mystery,
and find it made of opposites.
Opposites arise from the same source
and are identical in all but name.
The mystery of opposites is so profound,
to understand it
is to open the door of Tao.

LAO-TZU, TAO TE CHING

refused to participate in public life, and that he regarded his fellow citizens and the city's constitution with scorn. He eventually withdrew from Ephesus and went wandering into the mountains where he lived off wild plants. He only returned when he became ill and then died soon after of some incurable illness.

According to Diogenes, who later wrote of him, Heraclitus deliberately made his philosophical work obscure, so that none but those who really wanted to know themselves would be able to understand it. One of his works, which he entitled *On Nature*, he placed in the Temple of Artemis at Ephesus. The work has not survived, but fortunately there are quotations from it in the works of others. In what may be a fictitious story, Diogenes also recounts the time Euripides gave a copy of Heraclitus' book to Socrates to read. When asked his opinion of the book, Socrates replied, "The part I understand is excellent, and so too is, I dare say, the part I do not understand; but it needs a Delian diver to get to the bottom of it."

The divers of the island of Delos were known for swimming to tremendous depths to harvest sponges from the ocean floor, and if we follow the Delian divers down into the depths far enough to discern the outlines of Heraclitus' meaning, we discover a teaching that imparts nothing more than the imperative "know thyself" inscribed at the Delphic Oracle.

Listening to the Logos rather than to me, it is wise to agree that all things are in reality one thing and one thing only.

HERACLITUS, FRAGMENT 50

THE DELPHIC ORACLE

FROM THE DAYS of the Classical Greek thinkers the Western world has moved along a course in which knowledge has led to the dispelling of many mysteries and logic has pierced the veils of superstition. Discovery after scientific discovery has led us to understand what all the world's spiritual teachers have always known and taught: Truth is an inner phenomenon.

An inscription at the Delphic Oracle said: *Know thyself.*

The remains of the temple at Delphi (early 4th century B.C.E.), the site of the oracle

The origins of the Delphic Oracle are found in Greek mythology which tells of a battle between the Gods of the sky and those of Earth. The infant Apollo took control of the spot on which the temple stands by killing Python, the dragon snake that possessed it. He then took the form of a dolphin and swam out to sea to capture a group of sailors, and appointed them to be first priests of his cult. Apollo spoke through an oracle at the temple, an older woman of blameless life. She took the name Pythia and sat over an opening in the Earth where the body of the slain Python had fallen into a fissure. Fumes arising from its decomposing body intoxicated her into a trance, and allowed Apollo to possess her spirit. In this state she prophesied, speaking in riddles.

Know then thyself, presume not God to scan;
The proper study of mankind is man.
Placed on this isthmus of a middle state,
A being darkly wise and rudely great:
With too much knowledge for the skeptic side,
With too much weakness for the stoic's pride,
He hangs between; in doubt to act or rest;
In doubt to deem himself a god, or beast;
In doubt his mind or body to prefer;
Born but to die, and reas'ning but to err;
Alike in ignorance, his reason such,
Whether he thinks too little or too much;
Chaos of thought and passion, all confus'd;
Still by himself abus'd, or disabus'd;
Created half to rise, and half to fall;
Great lord of all things, yet a prey to all;
Sole judge of truth, in endless error hurl'd;
The glory, jest, and riddle of the world!

ALEXANDER POPE (1688–1744)
AN ESSAY ON MAN, EPISTLE II

One of the most ancient, pure and powerful yantras is the Shri yantra. The meaning of the word *yantra* is "instrument," and a yantra is something like our talisman or amulet. The Vedic culture, like many ancient cultures, had sacred geometric designs that were representative of their Gods and have sacred sound vibrations that corresponded to these designs.

The *Logos* of Heraclitus is what Lao-tzu in China called *tao*, and the Vedic sages called *rit*. The Sanskrit word *rit* or *rta* is derived from the root *ar*, meaning fit together properly, join correctly, suitably united. The term *rta* suggests a rhythmic or smoothly-turning, perfectly-balanced wheel that is related to the Greek *harmos* (harmony) and also to the Latin *ars*, that is the root of the English art and artist.

Creative Inspiration

Art, inspired by the *Logos*, is divine. The Vedic sages saw themselves as universal artists who, like the Gods they praised, put things together correctly and in the process helped create the reality they lived in. In the period roughly between 1000 and 700 B.C.E., when the earliest of their texts were compiled, they assumed the power to create new worlds through their sacred rituals. They chanted verses that held secrets of transformation and surrounded themselves with fields or spaces of deep and far reaching tranquility. Their contemplations have inspired a religiousness that has continued until the present and has inspired one of the world's most peaceful nations.

Aware perhaps that those who performed the sacred rites in the future might not always have the ability to intuitively discern the proper terms with which to describe important elements, the ancient mystics laid down certain principles in very practical terms. These sages, like the collectors of honey in the Sarmoun Darq, crafted detailed instructions into sacred texts and beautifully symbolic concepts, in forms that could survive long passages of time.

The masculine creative power, which was often known as the God Shiva, was envisioned in the form of a *lingam*, a sacred phallus represented by a pointed triangle whose point is upward-facing, while the feminine creative power, known as Shakti, took the form of the *yoni*, the sacred womb, depicted as a downward-facing one. The eternally creative combination of Shiva and Shakti was expressed in a mystical diagram called a *yantra*.

Meditation manuals from a much later time taught methods by which this and other *yantras* were to be envisioned deep within one's being. Accessing deeply embedded creative powers, meditators were taught to externalize their consciousness, or bring it forth into the world. This practice enabled them to identify so deeply with an external object that the dichotomy of subject and object dissolved. When this happened, and everything "fit together" into a harmony where opposites meet and disappear, where polarities do not exist, where all paradoxes are resolved and all contradictions disappear, these meditators would then have the internal power to transform the objective world by changing the inner subject. It became their task to bring the many Buddhas and Bodhisattvas, who were seen to live deep within their hearts, outward into the background of empty space—similar to the way a painter paints an image on a wall or canvas.

Although there are no texts from Ancient Egypt that explain their philosophy to us, we know the Egyptians had a profound wisdom that was deeply inspirational to the early Greek philosophers. We do not know in what terms they expressed what they knew or what techniques they practiced, but certainly the enormous monuments they built are based upon a deep internal rhythm and the ability to call upon their inner senses in order to manifest extraordinary works of art and architecture.

There is a Tibetan story told that once long ago one thousand princes gathered and vowed to become enlightened Buddhas. One among them was named Avalokiteshvara, and he promised to wait for his enlightenment until all the other princes had arrived.

For countless ages he watched and waited while many attained enlightenment. But for every one that attained, countless more entered the realms of attachment, greed, and bondage. Seeing this movement as an eternal situation, the prince was momentarily overcome by grief and, for a brief moment, lost faith. At that instant his body exploded into one thousand pieces.

Calling out to all the Buddhas for help, he was immediately pieced back together. His understanding deepened from the experience, and Avalokiteshvara rephrased his vow. He promised to wait until every single soul was free—knowing this would mean waiting forever.

SPIRAL SYMBOLS, JOURNEYS, AND DANCES

Bhairava (Shiva) is Lord of the Cosmic Dance and invokes followers' hearts to expand in a fashion similar to the nautilus shell he holds in his hand. This sacred shell is one of the instruments through which he initiates creation, and its Golden Spiral is more than a symbol of eternity; it is the actual "shape" of energy that is found in the anatomy of our own bodies.

The Pace of Yu from *Secret Essentials on Assembling the Perfected of the Most High for the Relief of the State and Deliverance of the People*

This diagram from a Chinese woodblock-printed book (12th century) describes a sacred spiral dance.

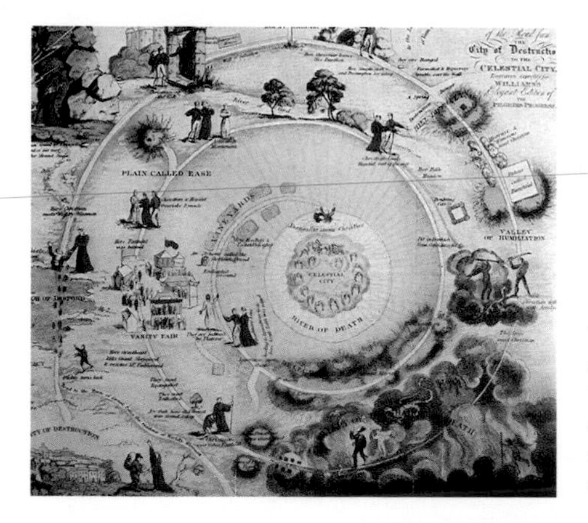

The circuitous journey described in John Bunyan's *The Pilgrim's Progress* is a spiral of unfolding consciousness in which similar situations occur on successive stages.

THE SUFI DANCE OF WHIRLING WAS INSPIRED BY JALALUDDIN RUMI.

There is a story told about a tradesman who sat in a shop in a small village. With his left hand he pulled a strand of wool from the bale that was above his head. He twirled the wool into a thicker strand and passed it to his right hand. The right hand wound the wool around a large spindle. The movements of the old man were continuous, and as he spindled the wool, he chanted la illaha illa'llah. There could be no uneven movement on his part or the wool would break and he would have to tie a knot and begin again.

The whirling dance of Sufis is a dance of divine love and mystical ecstasy. The music that accompanies the dancers has a melody played upon a stringed instrument that is accompanied by a voice. The words and even syllables of the poetry are connected to the melody. It is said that their music cannot be written in notes. Rumi said, "Notes do not include the soul of the dervish."

The dervishes (the name for these dancers) turn timelessly and effortlessly. They whirl, turning round on their own axis with their right hand turned up toward heaven to receive the divine energy that passes through the heart and is transmitted to Earth with the down-turned left hand. While one foot remains firmly on the ground, the other crosses it and propels the dancer round. The rising and falling of the right foot is kept constant by the inner rhythmic repetition of the name of Allah.

As waves upon my head the circling curl,
So in the sacred dance weave ye and whirl.
Dance then, O heart, a whirling circle be.
Burn in this flame—is not the candle He?

JALALUDDIN RUMI

A sculptured relief from Greece shows two dancers moving to the sound of a double flute.

Plato and His Thoughts on the Sacred Ratio

According to Plato in Book II of the *Laws*, the sacred ratios were studied by all those who became priests or practiced any of the arts that have a strong influence on society. Whether the sacred ratios that were studied refer specifically or more generally to the Divine Proportion we do not know, but it was through the study of them, he says, that early civilizations were kept stable and uncorrupted over thousands of years.

In a conversation that Plato describes, a philosopher from Athens speaks about the influence that music and dance have on the characters of young people and he refers to Egypt where, he says, certain laws were strictly upheld. In the time of Plato, the fourth century B.C.E., the Egyptians alone maintained a large body of knowledge concerning these divine ratios.

> *It appears that long ago it was decided by that nation to follow the rule of which we are speaking, that the movements and melodies executed by the young generation should be intrinsically good. The types to be permitted were defined in detail and were posted up in their temples. Painters and all other designers of dance or imagery were forbidden to deviate from these types or introduce new ones, either in their own arts or in music. The traditional forms were, and still are, maintained. If you go to look at their art you will find that the works of ten thousand years ago—I mean literally—are no better or worse than those of today, because both ancient and modern were based on the same standards.*

His listener asks the philosopher where these laws came from and the philosopher answers:

> *One could doubtless find things to criticize in other Egyptian institutions, but in the matter of music it is a very noteworthy fact that it has proved possible to canonize those forms of music which are naturally correct and establish them by law. This must originally have been the work of a god or a god-like being, and indeed the Egyptians attribute the forms of music which have been so long preserved to Isis.*

Apollo, God of the Sun, brings light. Crowned by the golden spirals of his rays that mimic the pattern of a sunflower, he represents an underlying regenerative principle of the universe.

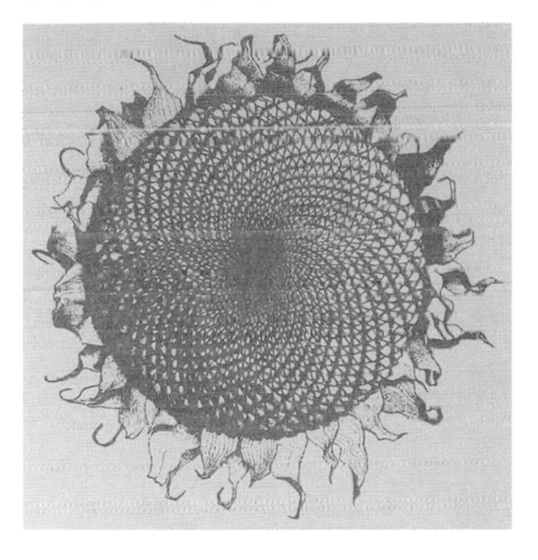

By Plato's time, the true forms of music had been forgotten everywhere except in the academies of Egypt where Plato had evidently studied them. The wane of theocratic rule and the rise of democracy had removed the restraints with which such laws were enforced, and the decline of ancient sacred music was later made complete by Christianity's opposition to the study of pagan science. So what was once a truly significant body of sacred knowledge has been lost to us except for the little that was preserved in the spirit of Greek understanding. After the great civilization of ancient Egypt and the wonderful spirit that emerged on the Greek mainland, there was a real dwindling of interest in the ancient wisdom and eternal truths; and the traditions that had kept them alive were replaced by a new and different fervor.

Experience of Oneness

The experience of unity in existence was fundamental to early Greek mystics. They made little or no distinction between the

In picturing divine love we understand from the deep embrace of lover and beloved that their hearts have merged and become one.

living spirit and the material world. Before Aristotle's theories of empirical logic, Heraclitus said that "all things are full of souls and spirits," and Plato later echoed this with, "all things are full of Gods." According to Aristotle, Thales of Miletus believed that "soul is intermingled in the whole universe."

Eastern experiences of unity have expressed the same truth again and again. Hussein ibn Mansur al-Hallaj (857–922) a Sufi mystic and martyr from Southern Persia, has said:

In that glory there is no "I" or "We" or "Thou."
"I," "We," "Thou," "He," and "She" are all one thing.

Jalaluddin Rumi (1207–1273), another Sufi mystic and poet, turns to love for his metaphor, where the beloved is God.

All the bright-plumed birds of heaven
Will devour their hearts with envy
In this place where we rest,
Thou and I.
It is the greatest wonder,
That we are sitting together here
In the same nook, at this moment,
Thou and I.

Fakhruddin Iraqi (13th century), another Sufi mystic:

"Whose beloved are you?" I asked,
"You who are so unbearably beautiful?"
"My own," he replied, "for I am one and alone
Love, lover, and beloved—
Mirror, beauty, eye."

Proportion, the Greek *Logos* that Heraclitus spoke of, was an expression of this intermingled symmetry—a harmony in which differences are unified, not, as they later became, split apart and polarized. Without the use of the mathematical logic we have today, the early Greeks understood proportion to be a relationship in which differences are part of the whole. *Logos* was to them an experience, not a concept; and it was a relatedness that included both observer and observed.

The inspired tranquility of this Egyptian blind harpist from the tomb of Paatenemheb (ca. 1330 B.C.E.) gives the impression that music is, in its essence, a direct response to the harmonics of proportion.

Logos and Music

What the Greeks knew intuitively about music was naturally reflected in the experience of *Logos*; for music *is* harmony. Harmony *is* music: It is the most natural expression of the mathematical laws the Pythagoreans discovered, and their mystical teachings attributed great significance to it.

There is a theory that early Christian mystics, when they referred to *Logos* or "the Word" were referring to music. We hear in the New Testament of the Bible that "In the beginning was the Word, and the Word was with God, and the Word was God." There are many ways to understand this beautiful statement. It is interesting to note that where we now read "Word" the Greek text has *Logos*.

> *All the planes of existence are involved in it. It is not only that the soul is full of music: once you have heard the inner music, your mind is full of it, your heart is full of it, your body is full of it, all your layers of being are full of it. Once known, not only do you hear it*

Macrocosm and Microcosm

Hindu cosmology speaks in the Upanishads of Brahma who has two sides–one of which is Existence, Consciousness, and Bliss (Satchitananda) and the other is the world we know (Maya), matter, life, and mind. These represent the upper and lower worlds of our being or the "Golden Egg."

Brahma created Maya in order to experience the delight of his own creation. Maya is the Mother of the World or the Great Womb that gives birth to the Golden Egg. When the Parabrahman enters into matter in the seed-state of things (Pakriti) and forgets himself, he still has a memory of himself that is deep within, in the Atman, which resides deep within him and just over his head.

As Parabrahman lives in the world of Maya and Illusion, he has a sense of his personal identity. He is conscious and has a distinct sense of his limitation. His goal is to unify all things within himself, and to do this he develops a conscious relationship with the divine, which measures out the exact proportion of those things that are necessary in the exchange of energy between this world and that. He always holds the key to free up the lower energies and raise them to the higher world.

By discovering his true nature and taking right action, he can merge with his higher bodies while here below. Through the right use of energy, he can scale the heights of the Bliss body and tap into the delight of being.

Human as microcosm of the macrocosm by Hildegard of Bingen (1098–1179)

Hildegard's painting describes the human body in the cosmos and the cosmos in the human body. One forms the other, she explained:

Now God has built the human form into the world structure, indeed even the cosmos, just as an artist would use a particular pattern in his or her work.

In her writings Hildegard describes this image:

I looked—and behold!—the east wind and the south wind, together with their sidewinds, set the firmament in motion with powerful gusts, causing the firmament to rotate around the earth from East to West.

Breath and spirit come from the same root word. Hildegard speaks of inhaling and exhaling the "breath of the world." She goes on to describe the universe as an organism, a single, breathing, living body.

inside of you, it is outside too. In the song of birds you hear it, and in the wind passing through the trees you hear it, and in the waves striking on the rocks you hear it. In sound you hear it, in silence you hear it.

In fact the greatest music in the world is nothing but an echo of the inner music.

OSHO (1931–1990), INDIAN MYSTIC

What motivated early mystics was a desire to come nearer to self-awareness and, thus, to God. Mathematics was a tool of language through which harmony could be expressed, and as these early mystics knew, divinity and harmony are one. To them God was unity, and how else but through proportion could creation reflect the creator?

What we discover, whether we look toward the macrocosm or toward the microcosm, is an existence in which everything is perfectly proportioned.

Division of oneness is impossible. We can neither grasp it in its totality, nor can we experience it in its complete complexity. We certainly cannot divide it. However, it is possible to get a sense of oneness when we perceive it through a sense of proportion. Looking at small things and contemplating them helps us to see bigger things more clearly. Watching the patterns in minutia illuminates as much as watching them in the larger worlds—and vice versa.

In the Hindu discription of things, Vishnu reclines upon a spiral shaped serpent, floating in the cosmic ocean within the Golden Egg. As he sleeps soundly, a lotus at his navel opens and Brahma emerges to live out his life of one hundred years. Vishnu, in his sleep, watches all. As each life of Brahma comes to an end, the lotus closes and Vishnu stops dreaming.

One year in Brahma's life lasts 360 days and nights. Each day is known as one *kalpa*, and it lasts 4,320,000 human years. Each night is as long as a day, but deeply silent.

During the passage of time there is a slow and irreversible deterioration of the cosmic order of things. Goodness vanishes by degrees from the world and humans become filled with lust and evil. When the world has deteriorated beyond any salvation, it is destroyed by Shiva.

Buddha

Knowledge

But looking and contemplating is not everything. We must at some point put aside certain kinds of knowledge and enter the world of experience. Mystics tell us that one of our greatest problems is knowledge: It can serve us and it can undo us. When we learn a thing by real experience we know it well, when we borrow some facsimile from the endless catacombs of other people's minds, we do ourselves the disservice of pretending to know.

There are beautiful Zen stories that tell of monks who approach their teachers with real heartfelt desires to experience true states of consciousness. Again and again they are tested. Then one day it happens. In a moment of unbelievable simplicity, the archer raises his bow, pulls back the string, and experiences the complete and empty void in which real being is known.

True inspiration floats out of these spaces. It is there that anything that is really worth knowing can be found. It is there that love resides and meditation flowers.

There is a story told in Buddhism of Gautama Buddha who wandered around India for forty years, speaking to groups of people who gathered to listen to his words. At one such gathering Buddha appeared with a flower in his hand. The deep hush that usually announced his arrival grew deeper, for on this day he did not begin to talk. He simply sat in silence with the flower in his hand.

As the silence grew more difficult for many, it grew more profound for others. Finally, from the back of the gathering a loud and heartfelt laugh rang out. Buddha looked over to where the laugh had come from, and as his eyes caught those of Mahakashyapa he smiled. Rising up from his seat, Gautama walked over to the laughing monk and handed him the flower, saying, "What can be said, I have already said. What cannot be said, has been conveyed to Mahakashyapa."

Islamic rose

The flower is a timeless symbol of truths that are mysteriously conveyed. The scent of its perfume carries an unspoken message, its transient beauty conveys a sense of the transitory phases of our lives, the flowering phenomena itself speaks of what is possible when we allow the divine to enter. The monks that sat in the Buddha's assembly were all taken aback. None had ever thought that Mahakashyapa was a man of any importance, none could see what made him laugh. No one will ever know for sure what transpired on that quiet morning, but many will guess.

Rose window at Chartres Cathedral

The mysterious conveyance of divine truths is something none of us can ever really comprehend. What happens in that perfect moment of openness and availability is something some people taste and most really only hear about.

I have seen my lord with the eye of my heart,
I said, "Who are You?" He said, "You."

HUSSEIN IBN MANSUR AL-HALLAJ

To love is to know me,
My innermost nature,
The truth that I am.

BHAGAVAD GITA (5TH CENTURY B.C.E.)

*We came whirling
out of nothingness
scattering stars
like dust.
The stars made a circle
and in the middle,
we dance.*

JALALUDDIN RUMI

LOOKING FOR TRUTH

One of the universal truisms about the mystical realms and the search for truth is that there is nowhere to go and nothing to find. The Taoist sage Lao-tzu (ca. 604–531 B.C.E.) said: "Without looking out of the window, one may see the way of heaven."

As we look at the Divine Proportion and try to tease out a code of universal importance, we really do not have to look very far. The intricate nooks and crannies that biologists, botanists, physicists, and mathematicians look into and describe for us are as beautiful as the edifices and works of art that painters, musicians, and architects create. What seems to lie at the root of the equation that nature scribbles and Euclid puts in formal terms is an acknowledgment of the unfathomable wonder we are a part of. There is an old Sufi saying that explains why God, wanting to experience what He had created, hid in the human heart.

It is the perfect vantage point, a unique hiding place. From there we can venture out toward the farthest stars or turn in and discover such similar worlds. We find similar shapes, similar principles, similar unknowns, and similar truths: A unique proportion guides us, one that says the Whole is in a perfect relation to its parts, for the Whole is to the larger as the larger is to the smaller—Divine Proportion.

Jacob's Ladder by William Blake

The door to the real world spirals through the heavens. Ascending and descending angels mark the spiral ascent of the human soul and the reciprocal descent of divine wisdom.

In ascending the spiral stairway of life we circle around our center, either moving closer to or further away from some understanding of who we are. Like segments of the Golden Spiral of Divine Proportion, each stair represents a shift in consciousness and demands a new awareness.

$$\Phi = \frac{(1+\sqrt{5})}{}$$

DIVINE PROPORTION Φ GO

POPORTION Φ GOLDEN SECT

RED CUT Φ DIVINE PROPOR

LDEN PROPORTION Φ GOLDE

SACRED CUT Φ DIVINE PRO

1.618033988749

DEN MEAN Φ GOLDEN PRO

ON Φ GOLDEN RATIO Φ SAC

ION Φ GOLDEN MEAN Φ GO

N SECTION Φ GOLDEN RATIO

PORTION Φ GOLDEN MEAN

3948482045868 3

GLOSSARY

Divine Proportion
The Divine Proportion is based on a relationship that says the whole can be compared to a larger part in exactly the same way that the larger part can be compared to a smaller one. Its mathematical expression gives us the number $(1 + \sqrt{5}) / 2$ or 1.61803....

Fibonacci numbers
The Fibonacci number sequence (0, 1, 1, 2, 3, 5, 8, 13, 21, 34, 55, 89, 144...) is one in which any number in it can be determined by adding the previous two numbers together. A graph that plots the ratios of each number in the sequence with the number that precedes it will approach the Divine Proportion.

Fractal
A fractal is a mathematically produced shape or behavior that describes phenomena such as lightning, trees, and clouds in ways that Euclidean geometry cannot. Fractals are generally self-similar, meaning that they repeat themselves on different scales within the same object. Fractals possess infinite detail.

Golden Ratio
The Golden Ratio is another name for the Divine Proportion, Golden Mean, or Golden Section. This was first stated mathematically by Euclid in *Elements*:

A straight line is said to have been cut in extreme and mean ratio when, as the whole line is to the greater segments, so is the greater to the lesser.

Golden Rectangle
A golden rectangle is a rectangle with dimensions that are of the Divine Proportion. When the measurement of one side is one unit, the other side will measure $(1 + \sqrt{5}) / 2$.

Golden Section
Golden Section is another name for the Divine Proportion and comes from the Latin *section divina*, which was first used by Luca Pacioli in his work *De divina proportione*, published in Venice in 1509.

Golden Spiral
A logarithmic spiral or equiangular spiral is a special kind of spiral curve that often appears in nature. The Golden Spiral is a particular version of this spiral that is based upon the Divine Proportion.

Golden Triangle
In geometry, the Golden Triangle is an isosceles triangle with a ratio of Φ between its side and its base. The two angles at the base of a Golden Triangle are 72 degrees. The angle at the apex is 36 degrees. This triangle has a unique property: It can be broken into two smaller triangles that are also Golden Triangles.

Integers
Integers are whole numbers including zero and the negative numbers: ... -3, -2, -1, 0, 1, 2, 3....

Irrational numbers
Irrational numbers are real numbers that cannot be expressed as a ratio of integers. Their decimal expansion does not terminate. They include $\sqrt{2}$, $\sqrt{3}$, and $\sqrt{5}$. Phi is an irrational number.

Phi
Phi (the Greek symbol Φ) is a mathematical symbol for the Divine Proportion. The decimal representation of Φ is 1.6180339887499....

Proportion
Proportion is a comparison of two ratios and includes an indication of how the two ratios are related.

Ratio
A ratio is an expression of the relationship between two quantities. The word comes from the Latin *ratio* meaning "to estimate."

Rational numbers
Rational numbers are a set of numbers that are written as the ratio of two whole numbers.

Sacred Cut
Sacred Cut describes the way in which a line or shape is cut so that the resulting parts are proportionate to Φ.

Sectio aurea
Sectio aurea is Latin for Golden Section.

Square root
The square root of a number is another number that can be multiplied by itself—or squared—to equal the number.

Tetraktys
The Tetraktys is a triangular figure consisting of ten points arranged in four rows: one, two, three, and four points in each row. As a mystical symbol, it was important to the Pythagoreans.

Selected Bibliography

Boles, Martha, and Rochelle Newman. *The Golden Relationship: Art, Math & Nature: Universal Patterns.* Bradford, MA: Pythagorean Press, 1990.

Conway, John H., and Richard K. Guy. *The Book of Numbers.* New York: Springer-Verlag, 1996.

Devlin, Keith. *The Language of Mathematics: Making the Invisible Visible.* New York: W. H. Freeman and Company, 1998.

Durando, Furio. *Ancient Greece: The Dawn of the Western World.* New York: Stewart, Tabori & Chang, 1997.

Duruy, Victor. *The World of the Greeks.* Translated by Joël Rosenthal. Geneva: Minerva, 1971.

Ghyka, Matila. *The Geometry of Art and Life.* New York: Dover Publications, 1977.

Gies, Joseph, and Frances Gies. *Leonardo of Pisa and the New Mathematics of the Middle Ages.* Gainsesville, GA: New Classics Library, 1969.

Guant, Bonnie. *Beginnings: The Sacred Design: A Search for Beginnings, and the Eloquent Design of Creation.* Kempton, IL: Adventures Unlimited, 2000.

Guedj, Denis. *Numbers the Universal Language.* New York: Harry N. Abrams, Inc., 1996.

Hawking, Stephen, ed. *The Illustrated On the Shoulders of Giants: The Great Works of Physics and Astronomy.* Philadelphia: Running Press, 2004.

Heilbron, J. L. *Geometry Civilized: History, Culture, and Technique.* Oxford: Clarendon Press, 1998.

Herz-Fischler, Roger. *A Mathematical History of the Golden Number.* Mineola, NY: Dover Publications, 1987.

Huntley, H. E. *The Divine Proportion: A Study in Mathematical Beauty.* New York: Dover Publications, 1970.

Ifrah, Georges. *The Universal History of Numbers: From Prehistory to the Invention of the Computer.* New York: John Wiley & Sons, 2000.

Kaku, Michio. *Hyperspace: A Scientific Odyssey Through Parallel Universes, Time Warps, and the Tenth Dimension.* New York: Doubleday, 1994.

Lemesurier, Peter. *Decoding the Great Pyramid.* Shaftesbury, England: Element, 1999.

Livio, Mario. *The Golden Ratio: The Story of Phi, the World's Most Astonishing Number.* New York: Broadway Books, 2002.

Mankiewicz, Richard. *The Story of Mathematics*. Princeton, NJ: Princeton University Press, 2000.

McDermott, Bridget. *Decoding Egyptian Hieroglyphs: How to Read the Secret Language of the Pharaohs*. San Francisco: Chronicle Books, 2001.

Michell, John. *The Dimensions of Paradise: The Proportions and Symbolic Numbers of Ancient Cosmology*. Kempton, IL: Adventures Unlimited, 2001.

Oakes, Lorna, and Lucia Gahlin. *Ancient Egypt: An Illustrated Reference to the Myths, Religions, Pyramids and Temples of the Land of the Pharaohs*. New York: Hermes House, 2002.

Pappas, Theoni. *More Joy of Mathematics: Exploring Mathematics All Around You*. New York: Wide World Publishing/Tetra, 1991.

———. *The Magic of Mathematics: Discovering the Spell of Mathematics*. New York: Wide World Publishing/Tetra, 1994.

———. *The Joy of Mathematics*. San Carlos, CA: Wide World Publishing/Tetra, 2004.

Purce, Jill. *The Mystic Spiral: Journey of the Soul*. New York: Thames & Hudson, 1974.

Rawson, Philip. *Tantra: The Indian Cult of Ecstasy*. New York: Thames & Hudson, 1973.

Schimmel, Annemarie. *The Mystery of Numbers*. New York: Oxford University Press, 1993.

Schneider, Michael S. *A Beginner's Guide to Constructing the Universe: The Mathematical Archetypes of Nature, Art, and Science*. New York: HarperPerennial, 1995.

Sigler, L. E. *Fibonacci's Liber Abaci: Leonardo Pisano's Book of Calculation*. New York: Springer-Verlag, 2003.

Struik, Dirk J. *A Concise History of Mathematics: Fourth Revised Edition*. New York, Dover Publications, 1987.

Sutton, Daud. *Platonic & Archimedian Solids: The Geometry of Space*. New York: Walker & Company, 2002.

Thompson, D'Arcy Wentworth. *On Growth and Form: The Complete Revised Edition*. New York: Dover Publications, 1992.

Toman, Rolf, ed. *The Art of Gothic: Architecture, Sculpture, Painting*. Cologne: Könemann Verlagsgesellschaft mbH.

Vitruvius. *Vitruvius: The Ten Books on Architecture*. Translated by Morris Morgan. New York: Dover Publications, 1960.

ART ACKNOWLEDGMENTS

Cover

Siede Preis

Introduction

2: Tate Gallery, London; 4 upper: NASA/Genesis; 4 lower: Lennart Nilsson; 5 upper: Dr. Johannes Durst/SPL; 5 middle: David Malin/Anglo-Australian Observatory; 5 lower: Michael Freeman; 6 upper: American Rose Society, Shreveport, LA; 6 lower: Dr. David Roberts/SPL; 7 upper: J.-C. Golvin; 8: Alexander Lieberman; 9: The Sistine Chapel, Vatican, Rome.

Chapter 1

10: Camille Flammarion, *L'atmosphère météroIogie populaire*; 11: The Art Archive; 12: Euclid, *Opera omnia* (1703), frontispiece; 13: Deir el-Medina, Theban Necropolis, Egypt; 14: Adana Archeological Museum, Adana, Turkey; 15 upper right: C. M. Dixon; 15 middle right: Anne Ophelia Dowden; 16 lower: Library of the Linnean Society, London; 18 upper: British Library, London; 18 lower: British Museum, London; 19: British Library, London: 24 upper: Windsor Castle, Royal Library; 24 lower: Metropolitan Museum of Art, NY; 26: Historiska Museet, Stockholm; 27: Matjuska Teja Krasek.

Chapter 2

28: British Museum, London; 29: Metropolitan Museum of Art, NY; 30 upper: Museum of Natural Sciences, Brussels; 30 lower: British Museum, London; 31 upper: Bibliotèque National, Paris; 31 lower left: Biblioteca Ambrosiana, Milan; 32 upper left: English Heritage Photographic Library; 32 upper right: The Schoyen Collection, Norway; 32 lower: British Museum, London; 33: Art Archive; 34 upper: Ancient Egypt Picture Library; 34 lower: Acropolis Museum, Athens; 35 left: National Archeological Museum, Naples; 35 right: Eric W. Weisstein; 36: Linean Society Library, London; 37: Bibliotèque National, Paris; 39: Metropolitan Museum of Art, NY; 40: Stanza della Signatura, Vatican Palace, Rome; 42: The Philosophical Research Society, Inc., LA; 43: Osterreichische Nationalbibliothek; 45 left: British Library, London; 45 upper middle: Art Resource; 45 lower middle: British Museum, London; 45 right: British Library, London; 46: Siena Cathedral, Siena; 47: Frank Spooner Pictures; 49: Osterreichische Nationalbibliothek, Vienna; 50: Musée du Louvre, Paris; 51 upper: Shelly and Donald Rubin Collection, NY; 51 lower: Museo di San Marco, Florence; 52 upper: Punjab Hills, India; 52 lower: Egyptian Museum, Cairo; 53 upper right: Tretyakov Gallery, Moscow; 53 lower right: Los Angeles County Museum of Art; 54 upper: Kunsthistorisches Museum, Vienna; 54 lower: Burgerbibliothek, Bern; 55 upper:

Gladys A. Reichard and F. J. Newcomb; 55 lower: James Morris; 56 upper: Michael Freeman; 56 middle: Anne Ophelia Dowden; 56 lower: Allen Rokach; 57 upper right: Metropolitan Museum of Art, NY; 57 lower right: Ian Warpole/Network Graphics, NY; 58 lower: National Museum, Athens; 59 upper: Archiv fur Kunst und Geschichte, Berlin; 59 lower: Eric Lessing Culture and Fine Arts Archives; 60 upper: Universitatsbibliothek, Heidelberg; 60 lower: National Gallery of Art, Washington D.C.; 61: National Trust, UK; 62: Louise Riswold Designs, Sausalito, CA.

Chapter 3

64: Lokman, *Shahanshahnama*, Istanbul, 1581-82; 65: Cheng Dawei, *Suanfa Tongzong*, 1592; 66 lower: British Museum, London; 67: Luke White: 69 upper left: Museé de Louvre, Paris; 69 lower left: Egyptian Museum, Cairo; 70 upper: National Gallery, London; 70 lower: Palazzo Pubblico, Siena; 72: British Library, London; 73: Edimedia, Paris; 74: Musée de la Civilisation Gallo-Romaine, Lyons; 75 left: Bibliotèque Nationale, Paris; 75 upper right: Museo Capitolino, Rome: 76 lower: Axiom; 77: Musée Guimet, Paris; 78: Library of the Topkapu Sarayi Muzesi, Istanbul; 79: Museum of Fine Arts, Boston; 80: Columbia University, New York; 83: Gregor Reisch, *Margarita Philosophica*, Freiburg, 1503; 86: Columbia University, NY; 87: Arts et Métiers, Paris; 89: Musée Cluny, Paris.

Chapter 4

90: Santa Maria della Grazie, Milan; 91: Georgia O'Keefe Museum, Santa Fe, New Mexico; 92: Galleria dell' Accademia, Venice; 93: Galleria dell Accademia, Florence; 94: Sunion, Mt. Olympus; 95: Bibliotèque Nationale, Paris; 96: Acropolis Museum, Athens; 97 both: David Finn; 98 upper left: Archaeological Museum of Athens; 98 lower left: Peter Clayton; 98 center: Giovanni Dagli Orti; 99 upper: British Museum, London; 99 lower: British Museum, London; 102 lower: New York Public Library; 103 upper and lower right: *Portfolio of Villard de Honnecourt*, Paris: Catala Frères, 1927; 104 upper: Musée du Louvre, Paris; 104 lower: Mary Evans Picture Library; 105 upper: Universitätsbibliothek, Heidelberg; 105 lower: Capodimonte Museum, Naples; 107 left: National Gallery, London; 107 upper center: Osterreichische Nationalbibliothek; 107 upper right; Smithsonian Institution, Washington D.C.; 107 lower right: Tate Gallery, London; 108 upper: Luca Pacioli, *Divina proportione*, 1509; 108 lower: Italy, 1994; 109: Luca Pacioli, *Divina proportione*, 1509; 110 left: Galleria Nazionale delle Marche, Urbino; 111 left: The Art Museum, Princeton University; 111 right: Albrecht Dürer, *Underweysung der Messung, mit dem Zirckel un Richtscheyt* Nuremberg, 1525; 112 upper: Private Collection; 112 lower: Musée du Louvre, Paris; 113 upper: Windsor Castle, Royal Library, London;

113 lower: Institute de France, Paris; 115: National Archaeological Museum, Athens; 116 upper: Biblioteca Apostolica Vaticana, Rome; 116 lower: Nanjing Museum, China; 117: Museum of Fine Arts, Boston; 118: Museo di San Marco, Florence; 119: Franchinus Gaffurius, *Theorica musicae*; 120: Delphi Museum, Delphi.

Chapter 5

122: British Museum, London; 123: Lindley Library of the Royal Horticultural Society, London 124 left: Frank Horvat; 124 right: Courtesy of General Atomics, San Diego; 125: Andrew Burbanks; 126 upper: Allen Rokach; 126 lower: Courtesy of IBM Research; 129 upper: Robert Robertson; 129 lower: NASA/Oxford Scientific Films; 130 upper: LennartNilsson/Albert Bonniers Forlag AB; 130 lower: Dennis Kunkel Microscopy, Inc.; 131 upper: David Furness/Wellcome Photo Library; 131 lower: Dennis Kunkel Microscopy, Inc., Kailua, Hawaii; 132: Leonid Zhukov and Alan H. Bar; 133 upper left: George Hayhurst; 133 lower left: W.M. Roquemore/Wright Laboratory, USAF; 133 upper right: NASA/SPL; 133 lower right: NASA/SPL; 135: Robert Galyean; 136 all: Harold Feinstein; 137 both: Lindley Library of the Royal Horticultural Society, London; 138 both: D.R. Fowler; 140 upper left: Dr. Alesk/Science Photo Library; 140 upper right: Museum of Modern Art, New York; 140 lower right: Tomb of Inheraku at Deir el-Medina, Egypt: 141 lower: Heather Angel.

Chapter 6

142: National Gallery of Art, Washington D.C.; 143: British Library, London; 144: Dr. Johannes Durst/SPL; 145 upper left: Biblioteca Marciana, Florence; 145 lower left: Kupferstich, 1742; upper right: District Museum, Torun; lower right: Nicholaus Copernicus, *De revolution ibus orbium socelestium libri VI*, Nürnberg, 1543; 146 upper left: Scala, Florence; 146 lower: Musée du Louvre, Paris; 146 upper right: Sternwarte Kremsmunster, Austria; 147 upper left: Farleigh House, Hampshire; 147 upper middle: S.F. Bause; 147 upper right: Photopresse, Zurich; 147 lower: Science Photo Library; 149 right: Lindley Library of the Royal Horticultural Society, London; 152: Museo Archeologia Nazionale, Naples; 153 left: Johann Kepler: *Mysterium Cosmographicum*, 1596; 153 right: Moonrunner Design; 154: Louise Riswold Designs; 155 upper: Courtesy of Paul Steinhardt, Princeton University; 155 lower: Collection Michael S. Sachs, Inc. Westport, CT; 156: Lindley Library of the Royal Horticultural Society, London; 158 left: Heather Angel; 159: Mel Erikson, Art and Publication Services and Ian Warpole, Network Graphics; 160 lower left: Al Sa'diyyin Tombs, Marrakesh, Morroco; A. Sonrel;

160 upper middle: A. Sonrel; 160 lower middle: Robert Harding; 160 right: M.C. Escher Foundation, Baarn; 161: Ian Warpole/Network Graphics; 163: Petra Gummelt.

Chapter 7

164: Punjab Hills, India; 165: Kahriye Camii, Constantinople; 166: The Ancient Art and Architecture Collection; 167 upper: Minaret of the Mosque of Samarra, Iraq; 167 lower: Sandro Botticelli; 168 middle. British Museum, London; 168 right: Zenrin-ji, Kyoto; 171: Metropolitan Museum of Art, NY; 172: The University of Chicago Library, East Asian Collection; 174: John Dugger and David Medala, London; 175: Museum of Fine Arts, Boston; 176 left: J. David; 176 upper right: Bibliotèque Nationale de France, Paris; 176 lower right: John Bunyan, *The Pilgrim's Progress*, 1678; 177 upper: Robert Harding; 177 lower: Metropolitan Museum of Art, NY; 178: British Museum, London; 179: M. Dixon; 180: British Library, London; 181: Rijksmuseum van Oudheden, Leiden; 182 upper: Victoria & Albert Museum, London; 182 lower: Hildegard von Bingen; 183: British Museum, London; 184: Private Collection; 185: Metropolitan Museum of Art, NY, 186: Galaxy Picture Library; 187: Biblioteca Nationale, Florence.

INDEX